Third Edition

LIVING IN WATER

an aquatic science curriculum for grades 5–7

Department of Education
National Aquarium in Baltimore

www.aqua.org

KENDALL/HUNT PUBLISHING COMPANY
4050 Westmark Drive   Dubuque, Iowa 52002

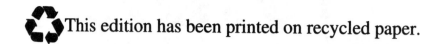 This edition has been printed on recycled paper.

NATIONAL AQUARIUM IN BALTIMORE

This project was supported, in part,
by the
National Science Foundation
Opinions expressed are those of the authors
and not necessarily those of the Foundation

The National Aquarium in Baltimore is not affiliated with the National Aquarium located in Washington, D.C.

CONTENTS

Air and water: gases and air pollution in solution

Soil and water: nitrate and turbidity

What are the characteristics of water with regard to temperature changes?
What are the consequences of temperature changes for the plants and animals
living in aquatic habitats?

Rates of change

INTRODUCTION

What is *Living in Water?*

Living in Water is a classroom-based, scientific study of water, aquatic environments and the plants and animals that live in water. It integrates basic physical, life and earth science. Mathematics and language arts are essential to its success. *Living in Water* is not a water monitoring program nor does it require access to an aquatic habitat although it includes suggested field experiences for students. It works with the field study of any body of water from the Pacific Ocean to a farm pond in Iowa. While many of its activities have been used in other projects, *Living in Water* is not a set of unrelated activities meant for infusion into existing curricula. As written, it is a complete, year-long curriculum which is carefully planned to build on its own foundations as it progresses through the year. With care, teachers or districts may select from among the activities to create shorter curricula. *Living in Water* has been adopted as formal science curriculum for both regular and special education students in a number of districts, ranging from large urban systems to isolated rural settings across the United States and in many other countries. *Living in Water* is self-contained. It is a teacher's manual with everything needed except the equipment.

What themes does it use?

Living in Water has several themes that run throughout its activities: control of variables in the design of valid experiments, the usefulness of models in understanding natural systems, application of knowledge in the design and testing of models, the collection and manipulation of numerical data, and identification of things using classification based on common characteristics. The years since *Living in Water* was first written have seen the development of the American Association for the Advancement of Science's Project 2061 and *Benchmarks for Science Literacy* followed by the National Academy of Science publication, *National Science Education Standards*. As basic, process-oriented science that incorporates the use of mathematics and language, *Living in Water* is consistent with many of the general recommendations of these important publications.

Who should use *Living in Water?*

It was originally written for grades four through six, but in the third edition we are recommending that it be used in grades five through seven. We believe it is more developmentally appropriate at this level, based on research about children's abilities to think and learn. This also matches the grades at which districts have tested and adopted *Living in Water*, with 6th grade being the most common. *Living in Water* is not appropriate for use below 5th grade. It can be modified for use at higher grades, including high school. Universities use *Living in Water* as a text in pre-service teacher education courses. It has formed the basis of intensive summer science enrichment programs. Students may find science fair projects among the activities and extensions.

How is it organized?

Living in Water is divided into six sections. An introduction to each section provides teacher information on science content, giving teachers the content they need to lead discussions following each exercise. Each introduction is followed by student exercises which place emphasis not on content, but on process. Students learn science content as they actively do science, reflect on the meaning of what they have done, write about it and discuss it in class, not from reading a textbook. Extensions and applications follow each activity.

How is it taught?

Various approaches and instructional strategies are used. In addition to experiments and classification activities, a number of exercises test the students' ability to apply basic principles to real problems through the development and testing of models. Students work in groups of varying sizes or alone, under the guidance of the teacher or independently. Varying the nature of instruction and the nature of the exercise accommodates different learning styles, making science accessible for all students, and models the variety of working conditions students will encounter as adults. Written reports generated comprise a portfolio of student work. Extensions enable students to pursue a variety of related topics. Some allow students to apply the results of their experiments to specific environmental problems. In other cases topics for independent study or library work are suggested. Most exercises include suggestions for using a classroom aquarium.

Where can I get help?

Supporting materials offer "how-to" information for preparation of materials and sources of supplies. Things like game boards, data sheets, reading pages and graph paper you can copy are included. While most activities have a worksheet, we strongly suggest using the worksheets as models for teacher-guided, student-designed worksheets. Some exercises have no worksheets and some require the enclosed worksheets which guide the students through independent work. Whenever possible, help the students design their own systems of recording data and examining results.

The review in this section, *The Inquiry Approach: Research and Trends* by Dr. Leon Ukens, will help teachers educate themselves, their principals and parents about why the emphasis in this curriculum is on process. The section *Reflective Thinking: Essential for Learning Science* by Sarah C. Duff, is particularly important for this kind of curriculum approach. Please read it carefully and incorporate its ideas as you teach each activity. For ideas about how to branch out on your own, giving the students more responsibility and control, see *Teaching Students to Be Scientists*. For updates, exchange of ideas and data with other teachers and students, and additional activity ideas, see the *Living in Water* home page (starting in early 1998) on the Aquarium's website at www.aqua.org. Still need help? Contact the editor at vchase@aqua.org.

I liked the old *Living in Water*. Why change it?

We didn't change what you liked. We've had 10 years of feedback from teachers. This third edition of *Living in Water* is modified and expanded. The largest addition is the inclusion of watershed issues.

Two additional water test kits were added (nitrate and pH) to enhance freshwater applications. Both affect biodiversity in many systems, and nitrate is a global change issue. Several activities were made more scientifically sound. Only one activity was deleted entirely. Many activities are streamlined. All are renumbered.

Funding for *Living in Water*

Living in Water was originally developed by the Department of Education and Interpretation of the National Aquarium in Baltimore under National Science Foundation grant no. MDR-8470190. Additional teacher institutes and part of this revision was supported by the National Science Foundation grant no. ESI 9254451. The National Aquarium in Baltimore has sustained this program with significant financial support since its inception.

Reproduction and use

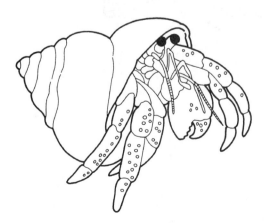

Participants and development

This project started with a three day meeting in July, 1985, of the authors, eight consulting teachers from the Mid-Atlantic states and a science educator. The consulting teachers were: Bonnie Bracey (Washington, D.C.), Cindy Dean (Delaware), Sarah Duff (Maryland), Margaret Gregory (Virginia), Jean McBean (Maryland), Jo Anne Moore (Maryland), Martin Tillett (Maryland) and Harold Wolf (Pennsylvania). Dr. Leon Ukens, Department of Physics, Towson State University, served as science education consultant. This group decided the nature and scope of the material. During the following school year, the authors at the Aquarium produced first and second drafts of thirty-six activities. In July, 1986, sixteen children tested the activities during a two week class at the Aquarium. Following revision, third drafts went to the consulting teachers, who tested them in their own classes during fall of 1986 and reviewed each activity. The third drafts were read for science education and science content by Dr. Leon Ukens and Gary Heath, of the Maryland State Department of Education. Dr. Thomas Malone, a biological oceanographer at the University of Maryland Horn Point Environmental Laboratories, and Dr. William S. Johnson, a marine ecologist at Goucher College, read the text for scientific content. During spring of 1987 the final draft was produced, using comments from the above persons and some 350 journals from students participating in the testing. First distribution was accomplished through two graduate courses for master teachers taught by the Aquarium authors at Goucher College during July, 1987.

In the first edition, Activities 4, 5 and 24 were written by Karen Aspinwall, Education Specialist, NAIB. Activities 21, 22 and 23 were done by Lee Anne Campbell, Education Specialist, NAIB. Activity 18 was written jointly by Lee Anne and Martha Nichols, Education Specialist, NAIB. Activity 16 was contributed by consulting teacher Martin Tillett, Howard B. Owens Science Center, Prince Georges County Public Schools. All other activities were the work of Dr. Valerie Chase, Staff Biologist, NAIB, who also served as editor and project director. Martha Nichols proofed final drafts and typeset copy for the entire curriculum. Layout, design and illustration are the work of Cindy Belcher, illustrator, NAIB.

The second edition, printed in 1989, was a modest revision based on comments from many sources, including the master teachers from the Goucher College courses. In 1995, *Living in Water* was again moderately revised for distribution on a CD-ROM.

This third edition is a significant rewrite by Dr. Valerie Chase and includes a number of activities she developed for Maryland teacher courses. It depended heavily on the advice, ideas and proofreading skills of Martha Nichols Regester, now Visitor Education Manager, NAIB, as well as Sylvia James, Director of Education Programs, NAIB; Bruce Barbarasch, Coordinator of Youth and Member Programs, NAIB (1995–1996) and Gretchen Linton, Education Specialist, NAIB. Invaluable additions and editorial review from Sarah Duff, Baltimore City Public Schools, helped shape many changes. Lisa Simpson and Associates did the layout and design with Cindy Belcher's illustrations.

The design, production and dissemination of this material is based upon work supported by the National Science Foundation under grants no. MDR 8470190 and ESI 9254451. However, any opinions, findings, conclusions and recommendations expressed in this publication are those of the authors and do not necessarily reflect the views of the National Science Foundation.

Valerie Chase, Ph. D.
National Aquarium in Baltimore
501 East Pratt Street
Baltimore, Maryland 21202
(410) 576-3887
vchase@aqua.org
May, 1997

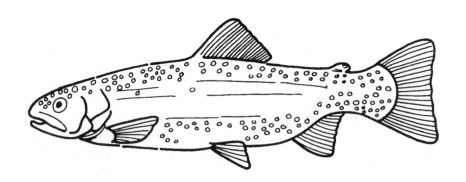

THE INQUIRY APPROACH: RESEARCH AND TRENDS

by Leon Ukens
Towson University
Towson, MD

The activities in this guide are based on a philosophy of science education in the upper elementary and middle school years supported by research studies of how children learn in those grades. This philosophy is one of teaching science by inquiry. The *National Science Education Standards* and Project 2061's *Benchmarks for Science Literacy*, two national publications with implications for the future of science education in the United States, also speak about the important role inquiry plays in doing science and speak positively on the role inquiry ought to play in science classes.

Applying the techniques of inquiry to answer questions posed by students, teachers, and curricula, is an integral part of our student's science education. (Benchmarks, 1993) Inquiry science has students involved in doing activities that eventually lead to conceptual development or change. These carefully developed hands-on activities, sometimes called guided inquiry, provide common experiences for students as they reflect back on what they already know and relate this knowledge to what they are learning. We want them to be able to recognize the relationship between explanation and evidence. (Standards, 1996)

Constructivism is a theory about how one comes to know something. While not a theory of teaching, it does speak to what should go on in the classroom. It requires concrete, meaningful experiences where the learner can search for patterns, raise questions, and construct meaning out of these experiences. The classroom becomes a community of learners engaged in whatever activity they are involved in. (Fosnot, 1996) Students should be able to construct their new conceptual understanding of the phenomena the activities are designed to develop. This involves a continuous process whereby moderately novel events cause learners to restructure what they know. If one compares how science is done to this style of teaching, a close correlation is observed. When scientists do science they make observations about a particular phenomenon of interest, develop models to explain these observations, and test these models. Models have two basic characteristics. They must account for the observations they purport to explain, and they should allow for predictions. This is precisely the way we think science should be taught. Students explore various phenomena. From these explorations and through reflection and discussion, concepts are developed. In order to understand the concepts, students should be able to apply them to new situations. In this kind of teaching, activities precede concepts. The need to learn something is developed by the materials in conjunction with the learner instead of coming only from the teacher.

But does inquiry work? In a nutshell the answer is YES! Inquiry science and constructivism are not new ideas. After Sputnik, a lot of effort was put into developing curricula which pursued science in an inquiry fashion. You probably know about some of these and perhaps even taught some of them. Three of the first were the Elementary Science Study (ESS), the Science Curriculum Improvement Study (SCIS), and Science, A Process Approach

(SAPA). Many research studies have been conducted in the years since their development in an attempt to find out how effective they were. Shymansky, et al. (1982) found that children in these "hands-on" science programs achieved more, liked science more, and improved their skills more than did students in non-inquiry classrooms. Bredderman (1982) in a similar study concluded that "with the use of activity-based science programs, teachers can expect substantially improved performance in science processes and science content, modestly improved attitudes toward science, and pronounced benefits for disadvantaged students." Curricula are still being developed at all grade levels that continue the teaching of science by inquiry.

The experiences in this guide follow a story line about "living in water" and all that is inherent in this rather broad topic. Conceptual understanding is what we care about—not merely recall of factual information. The role of the teacher changes in this type of classroom from one of information giver to one who facilitates not only the asking of questions, but also helps in pursuing answers. Questions like "How would you find out?", or "What is your evidence?" are often heard. (Martin, 1997) What constitutes doing good science and what constitutes doing good science teaching should be closely related.

Benchmarks for Science Literacy. 1993. Project 2061, American Association for the Advancement of Science, Oxford University Press, New York and Oxford.

Bredderman, Ted. 1982. Activity Science—the evidence shows it matters. *Science and Children*, Vol. 20, No. 1: 39–51.

Fosnot, Catherine. editor. 1996. *From Constructivism, Theory, Perspectives, and Practice*, Teachers College Press. Columbia University, New York and London.

Martin, David J. 1997. *Elementary Science Methods: A Constructivist Approach.* Delmar Publishing, Albany, NY.

National Science Education Standards. 1996. National Research Council, National Academy Press, Washington, D.C.

Shymansky, James, Kyle, William, and Alport, Jennifer. 1982. How effective were the hands-on science programs of yesterday? *Science and Children*, Vol. 20, No. 3.

REFLECTIVE THINKING: ESSENTIAL FOR LEARNING SCIENCE

by Sarah C. Duff, Supervisor
Specialized Support for Instructional Improvement
Baltimore City Public Schools

Something of great value is missing from many science classrooms across the country. As teachers have moved away from textbook-based science toward hands-on activities, they have not always brought with them the minds-on science that must accompany hands-on experiences if learning is to occur. John Dewey said, "We learn by doing, if we reflect on what we do." (Quoted in Marzano, 1992) But reflection is often missing from the science classroom.

Teachers who want to bring minds-on science into their classrooms, to help their students become more effective and reflective thinkers, will find three instructional strategies particularly useful.

1. Provide adequate time for thinking.

Mary Budd Rowe called science teachers' attention to the importance of "wait time"—perhaps better called "think time"—in 1974. Yet in too many classrooms teachers still use what Dr. Rowe called the "Inquisition" form of dialogue, posing a question, calling on the first student who raises a hand, and hurrying on to the teacher's next question.

Not only must think time be planned by the teacher, but the use of think time also needs to be taught explicitly: for example, the teacher might say, "Now I am going to ask you a question I want you to think about. I am going to give you the time you need to think about your answer. Please do not raise your hands until I signal you to do so." Then the teacher must insist that no one respond until adequate time for thinking has passed. After a student response, the teacher may pause again, encouraging the student to elaborate on his or her answer.

Enabling students to use think time is not easy at first. Students' experiences from their earliest years in school have taught them that the quickest answer, not necessarily the most thoughtful answer, is most desired by the teacher. Furthermore, the nervous teacher is sure that silence is an invitation to disorder in the classroom. Therefore, for the teacher who is newly focusing on improving student thinking, it is wise to begin with small increments of think time, moving gradually to longer time periods.

Teachers who teach think time explicitly and use it regularly find that more students participate and that answers become more thoughtful. Indeed, as students begin to appreciate the availability of think time and their own ability to construct thoughtful responses, they will demand think time if the teacher forgets to allow it.

2. Plan questions that require students to think at high levels. Listen to student answers.

In too many lesson plans, no questions appear, indicating that the teacher has given no advance consideration to the vital role questions play in stimulating student thinking. In too many classrooms, teachers' questions require students only to recall facts they have learned. But, if our students are to become proficient, reflective thinkers, teachers must plan with great care, formulating questions that will require students to use higher level thinking skills.

Higher level questions, at the appropriate developmental level, should be asked from the student's earliest years. Kindergartners who have taken a nature walk, for example, could be asked not only to name the things they saw (recalling), but such questions as

- How is your leaf different from Paul's? (comparing)
- Please arrange the objects you found from lightest to heaviest. (ordering)
- Find something longer than that stick. (comparing)
- How would the dirt be different if it were a rainy day? (drawing logical conclusions)
- How will the tree look in winter? (predicting)

Some educators reserve the teaching of higher level thinking until "the basics" have been learned. But what could be more basic than effective thinking? Acquisition of facts and using higher level thinking to reflect on those facts should happen together.

Teachers should use planned questions at every stage of instruction. When introducing a topic, teachers need to have ready questions that will help uncover students' misconceptions. Driver (1994) describes how a teacher, two weeks before he introduced the topic of rusting, gave each of his students an iron nail with the direction to put it in a place where it would get very rusty. Students were also to write answers to these questions:

- Where did you put your nail?
- What is it about that place which made you put it there?
- Why do you think that will make the nail go rusty?
- What do you think rust is?

From students' responses to these questions, the teacher gained a great deal of information about their thinking about rusting. Teachers should continue to use questions to check student understanding of the purpose of their activities. "Tell me why you are doing that," they will say repeatedly. They will use questions as embedded assessment, to check incremental knowledge gained. Teachers who have planned such questions will listen carefully to student responses and will use the responses in their subsequent planning.

Perhaps most important among a teacher's planned questions are those structured to help students reflect on the meaning of their hands-on experiences, the implications of their laboratory findings. Many teachers have discovered that they need to allow as much time for reflection on the meaning of an experiment as they allow for the experiment itself. They also need to plan the sequence of their questions carefully to lead the students toward achievement of the goal of the learning experience.

Some questions that will help students reflect on what their experiences mean are:

- Summarize your findings in one or two sentences.
- What do those numbers mean?
- Would it be useful to graph your findings? Justify your answer.
- Compare your findings with what you expected to find.
- What other findings would support your conclusions? Explain how.

- What findings would tend to negate your conclusions? Explain how.
- Identify possible reasons why Mary's results are different from Dante's.
- What if you had. . . .?
- Would you expect to get the same results if you conducted this experiment a year from now? Explain your answer.
- How is this model like the thing it represents? How is it different?

Thinking is often enhanced by having students write responses to such questions rather than responding to them orally. Questions should be both convergent (requiring students to work toward a single conclusion) and divergent (requiring students to develop many responses).

3. Engage students in metacognition.

David Perkins (1987), among other scholars, has found that just thinking at high levels does not in itself make students better thinkers. Only when they engage in metacognition, reflection on how they think, can students improve their thinking. Baker and Piburn (1997), writing about the importance of metacognition to construction of understanding, state: "Thinking about your own thought includes accessing and applying appropriate knowledge and strategies. Students with metacognitive skills can take charge of their own learning because they are aware of what they don't know and what they need to know for any given learning task."

The importance of understanding one's own thinking is also reflected in performance assessment programs like the Maryland School Performance Assessment Program, where students are often asked not only to solve a problem, but also to describe in detail how they thought in solving it.

"Tell me how you thought," is a request teachers need to make frequently, both when correct answers are given and when incorrect or unexpected answers are given. Responses to this question need to be followed up by probing questions that lead students to identify details of their thinking procedures. Some examples of such questions are:

- Tell me more about that.
- What did you think about first?
- How sure were you of your answer?
- What other problems have you solved that were like this one?
- What thinking strategies did you use to get ideas?
- How did you know you had finished this task?

Some educators have chosen to teach particular thinking skills directly and explicitly. (Beyer, 1987) In this model students engage in repeated use of a particular thinking skill and identify the thinking steps they use as they do so. An adaptation of this model is used in the Baltimore City Public Schools. For example, we have extended Activity 2 of *Living in Water* so that the thinking skill of classifying is taught explicitly in connection with the classification of aquatic habitats.

Other strategies are also important in helping students become lifelong reflective thinkers and learners: teaching students to see mistakes as opportunities; using cooperative learning strategies, and, perhaps most important of all, establishing the teacher's respect for and appreciation of each student. Furthermore, teaching the processes of thinking cannot be separated from teaching content. The richer and more extensive a thinker's content base, the richer and more significant his or her reflections can be.

Where will teachers find the time to enable students to become more reflective? Reform efforts such as Project 2061 sponsored by the American Association for the Advancement of Science (1990) point out that ". . . the schools do not need to be asked to teach more and more content, but rather to focus on what is essential to science literacy and to teach it more effectively." Similarly, the Third International Mathematics and Science Study, (1996) report describes how, in the United States "students cover more topics in mathematics and science than most of the other countries studied." "A mile wide and an inch deep" is the phrase the TIMSS report uses to describe U.S. curricula, textbooks and teaching. If some of the topics now taught in science are eliminated from the curriculum, perhaps students will be able to reflect on and construct more thorough understanding of those topics they do cover.

Living in Water is replete with opportunities for reflective thinking by students. As teachers use this curriculum, incorporating think time, asking the higher level questions, helping students articulate and manage their thinking procedures, reflection and understanding will be evident in their classrooms.

Baker, D.R. and M.D. Piburn. 1997. *Constructing Science in Middle and Secondary School Classrooms*. Allyn & Bacon: Needham Heights, MA.

Beyer, B.K. 1987. *Practical Strategies for the Teaching of Thinking*. Allyn and Bacon, Inc,: Boston.

Driver, R., A. Squires, P. Rushworth, and V. Wood Robinson. 1994. *Making Sense of Secondary Science*. Routledge: London.

Marzano, R., D. Pickering, A. Arredono, G. Blackburn, R. Brandt and C. Moffett. 1992. *Dimensions of Learning*. Association for Supervision and Curriculum Development: Alexandria, VA.

Perkins, D.N. 1987. Myth and method in teaching thinking. *Teaching Thinking and Problem Solving, 9(8)*.

Science for All Americans. 1990. Project 2061, American Association for the Advancement of Science. Oxford University Press.

Rowe, M.B. 1974. Wait-time and rewards as instructional variables, their influence on logic and fate control: Part one— wait-time. *Journal of Research in Science Teaching, 11* (2), 81–84.

Third International Mathematics and Science Study. 1996. U.S. National Research Center, Report no. 7.

TEACHING STUDENTS TO BE SCIENTISTS

Children are naturally curious and will "play" in an experimental way with materials you give them. Teaching them to experiment in a logical, scientific way without always telling them exactly what to do and how to do it is challenging. Experimental design can be developed out of group discussion. For the sake of saving space, this entire process is not repeated in detail in each exercise. You probably do not have time to do it for each exercise, but students should have the opportunity to control their own experiments sometimes. One process you might choose to use is:

- give the students a prompt and some equipment that stimulates thought about a specific area of knowledge
- have the students individually write one question that might be asked based on the prompt and the materials provided
- have the students quietly discuss what they have written in pairs
- have the students list the questions they think are possible on the board
- have the class choose one question and rewrite it for clarity
- have the class design an experiment so that it is a fair test—only one thing varies and all other parameters are controlled
- discuss what data will be collected and how; design a data collection sheet
- have the students do the experiment in groups
- refine the experimental design and/or data sheet as required during the exercise on the basis of consensus; discuss how they thought about these refinements
- share the data on the board or overhead projector
- have the groups do data analysis and conclusions independently in writing

- discuss how they thought about their work; what approaches did they use and how did they decide on them
- share conclusions and discuss them
- consider further directions for study
- what other questions are suggested by their work?

The single most common problem with experiments assigned by teachers is a lack of control of variables such that the difference between things tested cannot clearly be attributed to a single difference between the items studied. An example of a poorly designed popular project is to give students water contaminated with many chemicals and ask them to clean it up. This is an unfair test. By sheer trial and error, they may make some progress, but it is not an experiment. It may be a dozen or a hundred experiments. Real scientists would never approach this problem by trial and error the way students are forced to. They would run tests to identify the chemicals, and then experiment with the clean-up of known solutions of one chemical at a time.

Format and instructional strategies
ACTIVITY NUMBER and Title

The activity numbers are consecutive, from 1 to 50. The titles are sometimes silly, sometimes punny, more for you than your students.

Question and nature of the activity

The important part of the heading is the question(s) your students will try to answer in this exercise. For your convenience, we list the kind of activity this is.

Science skills

The skills listed are actions undertaken by students during this exercise. They are skills performed by scientists on a daily basis. They include:

- observing: using all five senses, taste is not appropriate in these exercises
- classifying: identifying like and unlike and grouping into sets
- measuring: using numbers to describe size, weight, quantity, volume or time
- organizing: analyzing and interpreting data, including the use of graphs, charts and tables
- inferring: drawing conclusions from data
- predicting: forming hypotheses based on past observations and results
- experimenting: identifying and controlling variables in testing hypotheses
- communicating: verbal, written, drawn or other forms of informing others about results and conclusions.

Concepts

The concepts listed are addressed by the exercise, generally proven by the activity the students perform.

Skills practiced

These are skills that students are assumed to be able to perform in order to do this activity. They are often taught in an earlier exercise, but some are skills assumed to have been taught in previous grades such as reading a thermometer. If activities are used out of sequence, particular attention should be paid to these skills.

Teacher's information

This is the content and context for the activity. It is given at the beginning for the teacher's convenience. It is not to be read to the students. Content should be developed with the students by the teacher after the exercise has been done. Words in CAPITAL LETTERS are science vocabulary that you and the students are expected to develop together during and after this exercise. They are defined in context and are often amplified in the content section that introduces each section. Each is also listed in the glossary. Words in **bold** are safety and environmental issues or very strong suggestions from the authors.

Introduction

This section suggests ways to begin to engage students in this activity. See *Teaching students to be scientists* for a more complete discussion on this and the following areas.

Action

Generally, you will start an activity by developing a class discussion. Take the time to draw out student understanding, identify misconceptions and build on previous class and personal experiences. The sequence of instruction often lends itself to an alternation of independent group work and class discussion.

Results and reflection

Have the students work up data and complete worksheets. Then lead a discussion. Students that have had a chance to discuss their ideas in a small group and then write them out are more likely to be able to articulate them before an entire class. Allow time for students to think. Engage them in thinking critically, using the section *Reflective Thinking* for details.

What to do if you ask a question and your students are unable to discuss it? If it is a question that you know they have the experience to answer, ask them to individually spend 5 minutes writing their thoughts. Then pick one other person and exchange ideas for 5 minutes. After this, return to a group discussion of the question.

Conclusions

This is the "take home" content information in this activity. Your students should master both the skills to do these activities and the content they develop.

Using your classroom aquarium

These are suggestions that relate to the content of this exercise.

Extension and applications

These are further investigations your students might undertake that follow the line of thought in this exercise. Or they are environmental applications of this content and skills learned here that students can explore independently or as a class.

Time

Estimated time is given in 45–55 minute class periods. The actual experiment can be done in less time, but students need to write, discuss, and examine their reasoning in order to reach closure on a lesson. "Less is more" from *Science for All Americans* refers to covering less content and spending more time reaching genuine understanding and mastery of what is covered. Extensions generally add to this time requirement. Teacher preparation is not included in this time estimate.

Modes of instruction

This curriculum does not make the assumption that there is only one best way to instruct students in science courses. Each activity includes a note on the mode of instruction. They vary from exercise to exercise for a variety of reasons that include student learning styles, safety, need for teacher control of behavior, degree of student mastery of techniques and equipment concerns. They also consider animal safety. Do **not** make teacher demonstrations into student activities. Categories of instructional modes include:

- teacher demonstration: the teacher does the exercise and the students observe; some students may be selected to make close observations or measurements for use by the class.
- teacher directed group work: where safety is an issue, the students work in groups while the teacher gives step-by-step instructions, the teacher keeps student attention and actions under control; most of the time, the students work in groups with the teacher giving introduction and occasionally stopping work to lead discussion; the rest of the time the teacher circulates among groups, helping students remain productive.
- independent individual, pair or group work—students work under the direction of a worksheet with little or no teacher input.

Sample objectives
These are behaviors your students should do in completing this activity. There are others that you may wish to add to this list.

Builds on
These are exercises that directly support this activity and provide preparation for its successful completion.

Materials
These are listed for the whole class, for each group or pair, or for each student with total numbers left to the teacher because each class has different numbers of students. Generally four students per group is the maximum, but sometimes larger groups are suggested. It is left up to the teacher to multiply the numbers of objects by the number of groups or students.

Under ideal conditions, student groups have trays of materials that they are responsible for keeping clean and in order. Plastic dish pans, plastic cat litter boxes or aluminum turkey roasting pans make inexpensive trays for equipment organization. Lists of items to be used can be written on the board as the class discusses the activity to be done.

Preparation

Work done by the teacher or aide prior to the class. Gathering the materials is assumed and not listed.

Outline

A brief listing of the sequence of actions recommended in this exercise.

National Aquarium in Baltimore

SECTION I Living in Water: aquatic habitats—freshwater, estuarine, and marine

Teacher's information

Children are naturally drawn to water. Whether they are playing in a pond, chasing waves at the beach or splashing in a rain puddle on a city street, children are entranced by water. The plants and animals that live in water and their ways of life are equally enticing. Children's natural interest is the motivation for *Living in Water*, which explores the nature of water, how it behaves in the natural world, and the plants and animals that live in it: their habitats, ways of life and interactions with each other. This exploration is largely through experimentation and hands-on activities. Two process themes run throughout the *Living in Water* activities: the nature of scientific experimentation with the need to control variables, and the nature and usefulness of models in science. These themes are introduced in this section and carried through the rest of the sections.

Section I also sets up the entire curriculum with two content themes: the physical characteristics of aquatic environments and their impact on aquatic plants and animals. Why do things live where they do? The physical characteristics of aquatic sites determine, in part, what lives there. Is the water salty or fresh, the bottom rocky or muddy, the water warm or cold, clear or muddy? In this section your students start to think about the physical characteristics of aquatic habitats that they will be studying throughout *Living in Water*.

The entire subsequent curriculum is based on scientific investigation of physical characteristics of aquatic habitats and their impact on living things. *Living in Water* is divided into sections that cover:

- things in solution and their impact on aquatic organisms and ecosystems (Section II)
- temperature, temperature changes and their impact on aquatic systems and organisms (Section III)
- viscosity, buoyancy and density of water and their effects on where and how aquatic organisms live and move through water (Section IV)
- the quality and quantity of light in aquatic systems and its impact on living things (Section V)

The final section (VI) of *Living in Water* is devoted to independent student investigation and to complex application problems that integrate information from across the sections of this work.

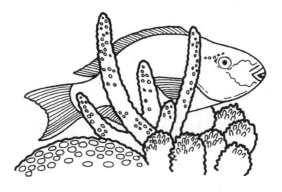

ACTIVITY 1 The disappearing act
What happens when different substances are added to water?
Student-directed experiments with known substances.

Science skills
observing
measuring
organizing
inferring
predicting
experimenting
communicating

Concepts
Many substances form a solution when mixed with water; some do not.

Some substances go into solution faster than others.

Skills practiced
bar graphing
measuring volumes of fluids and powders

Time
2–3 class periods

Mode of instruction
Independent student pair work

Sample objectives
Students design and conduct an experiment to compare rates at which different substances dissolve in water.

Students compare different factors which affect the rates at which some substances dissolve.

Students display data collected in a bar graph.

Teacher's information

This activity sets up everything that follows it. It has an essential function: to make students think about the nature of scientific investigation. Students must think about how to design a fair test within their own group. They must also consider that even though their group did a good experiment, they may not be able to compare their data with other groups because each group did something different. The entire rest of the curriculum depends on students standardizing their work so that they may compare their data.

Also, students begin to explore the characteristics of water which relate to forming solutions and suspensions. Table salt and sugar are both among the substances which form a SOLUTION with water: that is, they DISSOLVE, mixing completely with the water and staying mixed. Some substances appear to mix completely, but do not go into solution. When allowed to sit undisturbed, they settle out. These compounds are said to form a SUSPENSION. Corn starch is a household compound that forms a suspension with water.

National Aquarium in Baltimore

Students may not understand the difference between dissolving and melting and may use the two interchangeably. Melting is done by a pure substance when exposed to a higher temperature. Pure sugar melts when heated on the stove. Solutions are permanent mixtures of one or more substances in a liquid.

How you teach this experiment depends on the nature of your students. If this is their first experience with hands-on science, you may want to give them the experimental design with a clear purpose, explicit directions so that all the students are working on the same problems and the worksheets provided here. In this case you may want to discuss in detail with the students before beginning:

- why it is necessary to control VARIABLES
- why they must all use the same amount of water and chemicals
- why they must use the same techniques of stirring if they want to compare their results with each other.

If you are this explicit with the students, controlling the variable for them, then you should give them a second question in which they have to work together to design their test and control the variable for themselves.

If your students have some hands-on science experience, then you may teach this exactly the way it is written, using the data sheets to direct the student work and letting each group decide how much of each chemical to use. It also gives them some freedom to use different tools in stirring. This lack of standardization is intentional. Students cannot pool their data and compare their results. It generates vigorous student discussion about the need for uniform control of variables when doing an experiment as a class. This exercise is designed to produce arguments and discussion which require further experimentation to resolve. No amount of telling students about controlling variables is as valuable as letting them see what happens when you do not. Future exercises will be standardized for this reason.

Materials

for class
1 lb kosher or canning salt (see Recipes)
1 lb granulated table sugar
1 lb corn starch
24 clear plastic cups (about 6 oz)
24 stick-on labels or "post its"
1 qt canning jar or clear liter soda bottle with top cut off
1 pack dark flavor of unsweetened Kool-Aid

optional
rock salt
super fine sugar
sugar cubes

for each group of students
1/2 gal plastic milk jug or soda bottle of water
3 clear plastic glasses (6 or 9 oz)
3 stick-on labels
2 sizes of measuring spoons
a plastic teaspoon
a graduated measuring cup

for each student
work sheet
graph paper (see Recipes)

Preparation

Divide the "chemicals" into sets for the students. They should work in pairs, but several groups at a table may share materials so 8 sets of materials should be enough for the whole class. Label 8 cups "salt," 8 "corn starch" and 8 "sugar." Put about 1/4 cup of the chemical in each cup. If you are adding extra variables, the "sugar" cups may have superfine, regular or cubed sugar. The "salt" cups may be rock salt (ice cream salt), canning salt or kosher salt. Do not use table salt meant for salt shakers as it has calcium chloride added to keep it from clumping. This makes it cloudy in solution. Copy the data sheets.

Outline

before class

1. prepare materials and worksheets

during class

1. do the demonstration, pouring Kool-Aid very slowly into water after students predict outcome

2. introduce the words DISSOLVE and SOLUTION

3. show the students the materials

4. challenge them to design their own test to find out which go into solution and how fast, using the number of times stirred as a measurement

5. help students while they work in pairs with worksheets

6. review and discuss results

7. have them design a valid test, controlling VARIABLES

8. repeat new test and discuss

9. check corn starch following day

Teachers with particularly thoughtful students may want to encourage even more discussion by adding another set of variables besides the amount of substance used. Some groups may be given kosher salt, some fine canning salt and some rock salt (all NaCl but with different crystal sizes). Superfine and regular sugar also have different crystal sizes. Sugar cubes add one more point of confusion. The outcome is that different groups will get different results. Making them identify reasons for the differences will clearly demonstrate the need for standardized experiments in future activities in order to pool data.

Teachers of highly experienced students may direct their students to ask their own question about things in solution, list the exact steps they will take to answer it, list the equipment and materials they are using, design their own data sheet, compare their results, and compare their data sheet to the one given when they complete their work.

Introduction

To spark students' interest, show them a large clear jar of water and a package of unsweetened Kool-Aid. Ask them to predict what will happen if you pour the Koolaid into the water. Do they all agree? Slowly pour the Kool-Aid in and see what happens. It should sink and then begin to dissolve and spread (diffuse) through the water. Can the students suggest a way to speed up the process of dissolving? Stirring is one approach. A second would be to use hot water. Introduce the word SOLUTION for the mixture and the word DISSOLVE for the process of mixing completely. Clarify the difference between dissolving and melting. Do not stir the jar. Let it sit and see what happens over time. If you have an old, lumpy opened package of Kool-Aid, it takes several days to mix which adds interest. Unfortunately, a new, fresh pack mixes fast.

Ask what other things the students can name which they would find around the house that might dissolve in water. List them. Show the students the salt, sugar and corn starch. Ask them to predict whether each will go into solution. If they predict that these substances go into solution, could they design a test of their prediction? Second, could the students design a test to discover which dissolves fastest? Write their questions and predictions on the board.

Action

Give each pair the materials listed and the worksheet. Let them work independently without instruction from you. They will do the following as they use the directions on the worksheet. Each pair will label one clear plastic cup each with "salt," "sugar" and "corn starch." Fill each of the three plastic cups with about the same volume of water at room temperature. Leave about 1 inch of space at the top. Add equal amounts of salt, sugar and corn starch to the cup of water labeled with that substance. For example, 2 heaping teaspoons to each. Observe what happens for two minutes. Then stir each cup by making a circle around the edge of the cup with the plastic spoon ten times. Was there a change? Repeat stirring ten times in each until one has completely disappeared or dissolved. Record how many times it was stirred. Continue stirring and observing the other two to find out which dissolves next fastest. Last?

In order for this to be a "fair test" each cup has to be treated exactly the same way. The only thing that can be different is the substance added to each. This is referred to as controlling variables.

Results and reflection

Each group should have three numbers which are the number of times each substance was stirred before it dissolved. How do you display such information? A bar graph would be a good way to compare three different things.

Ask the students to compare their results with the predictions. Were there any surprises? Sugar or salt may be faster depending on the size of the crystals in the particular brand you buy. If you intentionally gave them different crystal sizes, let them compare each other's chemicals to see if they can identify reasons for different outcomes.

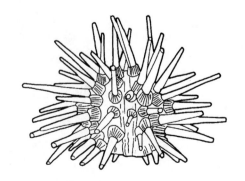

Did they all agree? Did the fact that different groups used different amounts of chemicals or water make a difference? Did they use the same stirring techniques? Can they fairly compare their results? Ask the class to help you write a list of all the possible VARIABLES (things that were different about how they did the tests) among their groups. Can you all get the same answer to the question if each group did a different experiment? They should get different numbers of stirs because they did not do the same tests even though they followed the same worksheet. Each group controlled variables within their group, but the class did not control variables among groups.

Explain that in the future, their goal will be to control variables uniformly so that they can compare their results fairly among groups and pool their data for analysis. In other words, in future experiments they will have to agree on uniform procedures before starting.

What is the single thing that can be different about each cup if this is to be a good experiment? The kind of chemical. Everything else must be the same. Have the class work as a group to restate the questions and then standardize experimental procedures. Have groups repeat the test. Now do they agree?

What happened to the corn starch can be the subject of heated debate. Some will say it is in solution, and others will not. Do not throw all the solutions away when students clean up. Save two sets of solutions and place them in a safe place overnight. When you check them a day later, you will have proof that the sugar and salt are completely mixed and are in solution while the corn starch was in suspension and has settled out.

Conclusions

What conclusions can your students draw from their results? There are three:

- in a fair test to answer a question, there can be only one difference between the things compared
- in order to compare data among the students, they must standardize their testing so that all do the same experiment

- some, but not all substances, go into solution in water.

Students should also be able to observe that some things dissolve faster than others. They may also be able to conclude that crystal size was inversely related to speed of dissolution.

Extension and applications

1. Can your students design an experiment to test whether things go into solution faster in hot water? They should be able to state the question, design an experiment in which there is a test (hot water) and a CONTROL for comparison (room temperature water) and carry out their test.

2. Demonstrate how to get substances out of solution. Use very hot water to make a very concentrated solution of salt or sugar. Put the cooled solution into a jar and suspend a nail or nut tied to a string suspended from a stick. Put the solution in a warm, dry spot and check them daily. Water EVAPORATES, leaving the dissolved substance behind.

Activity 1
The disappearing act

A. Write this section working alone

1. You and your partner are going to be doing an experiment. You are going to add salt, sugar and corn starch to cups of water. Based on what you know about these household chemicals, predict (guess) what will happen to each substance when it is added to water.

   Salt

   Sugar

   Corn starch

B. Working with your partner

1. Fill three clear plastic cups with about the same amount of water. Leave about an inch at the top so you can stir. Label each so you know which gets salt, sugar and corn starch.

2. Decide how many teaspoons of chemical you will add to each cup.

   Circle 1 2 or 3. Each cup gets the same number.

3. Add the substances, sit and observe for two minutes. Describe what you see happening.

4. Stir each cup 10 times using the same technique. Record the stirs. Did any dissolve?

5. Repeat stirring each 10 times and then observing it. Mark out the number of times each was stirred before the substance disappeared (if it did so).

    Salt

    10 20 30 40 50 60 70 80 90 100 110 120 130 140 150 160 170 180 190 200

    Sugar

    10 20 30 40 50 60 70 80 90 100 110 120 130 140 150 160 170 180 190 200

    Corn starch

    10 20 30 40 50 60 70 80 90 100 110 120 130 140 150 160 170 180 190 200

6. Using the graph paper, make a bar graph of showing your results.

C. Discuss this with your partner and then write the answer in your own words.

1. Which substances dissolved?

2. Which dissolved fastest? Give evidence for your statement.

3. Write a complete sentence that is one conclusion that you could make based on the results of your experiment.

Activity 1
The disappearing act

A. Write this section working alone

1. You and your partner are going to be doing an experiment. You are going to add salt, sugar and corn starch to cups of water. Based on what you know about these household chemicals, predict (guess) what will happen to each substance when it is added to water.

   Salt   Salt will go away so it does not show. Ocean water has salt and it is clear.

   Sugar   Sugar will probably mix. It will make the water taste sweet. It might not mix completely. Sugar is left in the bottom of the glass sometimes with iced tea.

   Corn starch   This does not look like it would mix easily but I have never used it.

B. Working with your partner

1. Fill three clear plastic cups with about the same amount of water. Leave about an inch at the top so you can stir. Label each so you know which gets salt, sugar and corn starch.

2. Decide how many teaspoons of chemical you will add to each cup.

   Circle 1 ②or 3. Each cup gets the same number.

3. Add the substances, sit and observe for two minutes. Describe what you see happening.

   The salt and the sugar fell to the bottom and spread out. Some funny lines formed above them. The cornstarch stuck together. Some stayed on top and some sank.

4. Stir each cup 10 times using the same technique. Record the stirs. Did any dissolve?

   There seems to be less of the sugar and the salt. The corn starch got milky and mixed in.

5. Repeat stirring each 10 times and then observing it. Mark out the number of times each was stirred before the substance disappeared (if it did so).

Salt

~~10~~ ~~20~~ ~~30~~ ~~40~~ ~~50~~ ~~60~~ ~~70~~ 80 90 100 110 120 130 140 150 160 170 180 190 200

Sugar

~~10~~ ~~20~~ ~~30~~ ~~40~~ ~~50~~ ~~60~~ 70 80 90 100 110 120 130 140 150 160 170 180 190 200

Corn starch

~~10~~ 20 30 40 50 60 70 80 90 100 110 120 130 140 150 160 170 180 190 200

6. Using the graph paper, make a bar graph of showing your results.

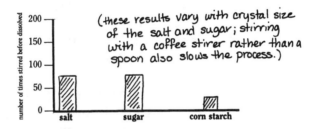

(these results vary with crystal size of the salt and sugar; stirring with a coffee stirrer rather than a spoon also slows the process.)

C. Discuss this with your partner and then write the answer in your own words.

1. Which substances dissolved? We both think salt and sugar dissolved but we don't agree about corn starch. I don't think it dissolved because it is milky.

2. Which dissolved fastest? Give evidence for your statement. Sugar dissolved a little bit faster. We stirred it less before it disappeared — 10 times less. Salt took 70 stirs. Other groups did not get the same results.

3. Write a complete sentence that is one conclusion that you could make based on the results of your experiment.

Sugar and salt disappeared and dissolved, but corn starch settled to the bottom of the cup.

(Teachers: the results vary with crystal size of chemicals and student techniques.)

ACTIVITY 2 Water, water everywhere

A classification exercise that defines water environments by physical characteristics as it teaches the use of keys and the nature of common aquatic habitats. Independent student work.

Science skills

classifying

Concepts

Observed characteristics allow you to identify something and to place it in a system that groups similar things.

Aquatic habitats differ from each other based, in part, on their physical characteristics.

Skills practiced

following a flow chart
using a key
reading for information

Time

1–2 class periods

Mode of instruction

independent individual or student pair work

Sample objectives

Students classify different kinds of aquatic habitats.

Students use a flow chart and/or a scientific key.

Students become familiar with the diversity of aquatic habitats.

Builds on

Activity 1

Teacher's information

This activity has three functions:

- it introduces students to aquatic habitats
- it allows them to practice using a key
- it introduces the physical characteristics of aquatic habitats which are the basis of all of the following sections.

Make sure you place emphasis on these physical characteristics in the review portion of your lesson. It serves as both introduction and motivation for the entire curriculum.

When humans group and name things, they are CLASSIFYING them. Things that are classified are first named and then placed in larger groups of things that share similar characteristics. For example, many kinds of tables are lumped under the term "table" as are many kinds of chairs under "chairs." Both tables and chairs belong to a larger category, furniture. The inclusion in ever larger groups results in a hierarchical organization of groups.

Why do humans bother with classifying things? Classification requires that we look for relationships among things, an activity that enhances our understanding of their functions and characters. Also, knowing that something belongs to a certain group means that you know something about it if you are familiar with the characteristics of the group. Finally, naming things helps us discuss them with each other and to search for information about them in the library or on the Internet.

This activity introduces students to habitat classification based on the physical characteristics and kinds of plants that grow in them. Students discover for themselves the different kinds of places that marine and aquatic organisms live rather than sitting and listening to a lecture, movie or video. The activity requires that students practice reading for information when they look for clues in the habitat description.

Students use a FLOW CHART to visualize the process of classification. At each step they must choose between two characters in order to proceed to the next step. Older students may also use a scientific KEY which is the same thing as the flow chart, but organized in a different format. As they do the activity, your students will also teach themselves about these habitats.

In order to use the key, your students will need to understand the way several words are used: salty for water that has an unknown amount of salt, salt water for water that is the saltiness of ocean water around the world, fresh water for water that does not have much salt nor taste of salt, and brackish for water which is between ocean water and fresh water in saltiness.

Please note that there are many other habitats that could be added or subdivisions of habitats made. For example, mud flats were not included. In warm, dry places like southern California, salt marshes include some areas that actually get saltier than salt water during hot dry summer periods. These were also deleted for the sake of clarity.

Introduction

Classification is based on the CHARACTERISTICS things have. Water environments have specific characteristics. Ask the students what kind of water is in the ocean. Remind them of Activity 1 in which some chemicals went into solution in water. The chemicals in solution in water are one characteristic of water environments. What comes out of the tap? Make sure that the students know the terms SALT WATER, FRESH WATER and BRACKISH WATER. Salt water has the salinity of the oceans, fresh water has little or no salt (you cannot taste any) and brackish water is a mix of salty ocean water and fresh water so it tastes less salty than the ocean. One misconception of many people is that only table salt is in the ocean. Sea water actually has many kinds of salts.

Materials

for class
habitat flow chart on an overhead
magazines
construction paper
optional
salt water (35 gm or 3 tbs table salt per liter or quart of water; see Recipes)
tap water
brackish water (mixture of tap water and salt water for brackish water)
small disposable paper cups
for individual or pair of students
habitat card
flow chart
key

Preparation

Copy the flow chart and key. Make habitat cards if this is your first time using this activity. You may simply duplicate the cards at the end of this section and use them as is. For nicer cards, glue them to one side of stiff colored paper. Your students may color the cards or, better yet, you may glue photographs of the same habitat cut from magazines to the reverse side from the habitat description. If you have students who have never traveled outside of an urban area, it is important to have large cards (9" x 12" construction paper) which have lots of pictures of the habitat described on the reverse side. Laminate the cards after the pictures are attached.

Ask parents to donate magazines from the Nature Conservancy, National Wildlife Federation, National Geographic, or other natural history magazines. Another source of magazines may be your public library, Goodwill or other organization that takes magazine donations and then sells them for a modest fee. Look for articles from the Everglades (for mangroves and seagrass beds) and Utah (salt lakes) particularly as these are hard to find. Aides or students may cut out pictures. Laminated cards to last for years. Making these the first time is hard work, but they look great and give students more inspiration. You might not find pictures of all of the habitats—it's all right to leave some out. Make duplicates of the most common ones.

Outline

before class

1. prepare habitat cards

2. make water samples for tasting

3. copy flow chart and key

during class

1. introduce the terms SALT WATER, FRESH WATER, and BRACKISH WATER; optional tasting

2. pass out flow chart and have pair demonstrate its use with one card

3. repeat with key

4. pass out cards and let students work independently, trading cards to complete chart

5. review and have students list characteristics on board

You can have your students taste solutions that taste like the ocean. Then compare with fresh water. A sip of salty water made with table salt will not hurt though it does taste bad. To make your point about brackish water, mix the fresh water with the salt water while they watch. Salinity is measured in weight of salt per total weight of the water. Salt water (ocean or sea water) is 35 grams of salt per 1000 grams of salt water. This is expressed as parts per thousand. Salt water is 35 parts salt per 1000 parts salt water or 35 ppt. Fresh water is less than 0.5 ppt and brackish water is in between. Salt water is written as two words when salt is an adjective modifying the noun, water. It is one word, saltwater, when the whole word is an adjective modifying another noun, such as saltwater habitat. The same rule applies for fresh water, rain forest, tide pool, etc.

Now what does the word HABITAT mean? Let the students discuss this. The concept includes all the things that make up an animal or plant's home. It is the place where a plant or animal normally lives and is defined, in part, by a set of physical characteristics such as salinity and, in part, by the kinds of dominant plants. Can they name any AQUATIC places (water habitats) where plants and animals live? Write their suggestions on the board. Pass out the habitat cards for them to examine. How can we identify an aquatic habitat that we have never seen before? We can use a CLASSIFICATION system based on characteristics which are written in clues on the back of the cards.

A KEY or FLOW CHART shows groupings that share common characteristics, arranged in large groups that keep getting divided into smaller ones. Use students to demonstrate the use of the flow chart first. Put it on an overhead projector and have one student read a card aloud while the second demonstrates that at each stage they must make a choice between two things until they come to a group in which all the things have the same characteristics and cannot be divided further. This is the name of their aquatic habitat. Explain that when identifying things using a system of classification, one starts with the biggest category and begins to work down to small groups. Here the first category is aquatic habitats, and the characteristic they have in common is that they all are in water. For younger students, you may choose to only use the flow chart. Older students may also use the KEY which may also need to be demonstrated. It is identical to the flow chart, but uses numbers to lead you rather than lines. At each step there is always a choice between two things. Note: if the card says brackish or salt water, go with brackish water in the key.

Now each student or student pair is going to become a mystery aquatic habitat. Turn the cards over to the description and pass out keys. Each will discover what he/she is by following a classification system that divides habitats up by their characteristics. Students will discover their characteristics or traits by critically reading for information from habitat cards. These cards have the clues to their identify.

Action

Turn the students loose. Trade cards until they have seen each one. As each habitat is identified, students may make a check mark next to it on their flow chart. Let students work independently until they finish.

Results and reflection

Use class discussion to pull ideas together. Pick several cards and ask students what they were. Ask the students to describe the reasoning that led them to their conclusions. Which information in the clues was most useful? What mistakes did they make in reading the clues? The most common is that they expect the clues to be in the same sequence as the key. They needed to read the entire description to find all the clues first. Allow students to check their results themselves by putting an answer key on the overhead. As a group, go over any that were confusing and rekey them on the overhead. If there is a disagreement, let the students resolve their differences by citing evidence.

Based on what they learned, which habitat would they like to learn more about and why? Why might we need to learn about different aquatic habitats? What might be the value of classifying them? Have you ever found it useful to classify something? Why might a scientist classify something?

Answer key

1. ocean or sea
2. sandy beach
3. rocky intertidal
4. continental shelf
5. kelp forest
6. coral reef
7. estuary
8. sea grass bed
9. salt marsh
10. mangrove swamp
11. river
12. stream or creek
13. lake
14. pond
15. freshwater marsh
16. swamp
17. bog
18. salt lake or salt pond

Conclusion

Have the class make a list of the physical characteristics used to classify aquatic environments. These are some of them:

- salt, brackish or fresh water
- warm all year or has seasons (tropical or temperate)
- flowing or still water
- clear and lighted or murky and dark
- large or small size
- tides or currents or waves
- shallow or deep
- sandy, rocky or muddy bottom
- plants submerged or sticking out of the water
- near shore or away from land

Make the point that some of these are going to be studied in detail during the weeks to come.

Using your classroom aquarium

As an assessment, have the students answer these questions about the classroom aquarium:

- Which aquatic habitat most resembles your classroom aquarium? Give evidence for your answer.
- Does it have fresh or salt water?
- What kind of bottom does it have?
- Does the water flow or stand still?
- What is its temperature?
- Try keying it out. The gravel on the bottom may be a problem.

Extension and applications

1. Ask the students to write one page about the habitat they would most like to visit and why. What would they imagine themselves doing there? What would they expect to see, smell and feel there?

2. Students may test their own knowledge of aquatic habitats following this activity. Make a set of cards, each with a loop of string long enough to hang behind the students' backs. Write the name of an aquatic habitat on each card. Hang them behind the students.

Students must find out what kind of water habitat they are by asking other students only yes/no questions about themselves. What strategies did they used in questioning? The best is to start with the big categories just as the flow chart did. You can equalize the chances by making sure the best students get the hardest ones.

3. Test student understanding of the principles governing classification and the construction of keys. Have students classify groups of other things and make their own key. The actual construction of classification systems is a higher level thinking skill. Require the students to justify their choices in their classification system. Creative choices of things might include keys to different groups of adventure toys, model or figurine collections or rock or rap groups. Let them trade keys to test the quality of their work. There should be at least ten items in each key.

Aquatic Habitat Cards

3. Your rocky shore is covered with seaweeds that live attached to the rocks. When your salt water (35 ppt) uncovers the rocks at low tide, sun, snow or rain falls on your seaweeds and animals. Waves crash into you, so animals and plants have ways of clinging tightly to your rocks.

You are _____

4. Your sandy or muddy bottom is always under salt water (35 ppt). In some places the water is deep, but you are along the edge of land. Animals burrow in your sand or mud. Your water is rich in tiny plants which provide food for many animals. Fishermen harvest your fish and shell-fish.

You are _____

1. You are a big body of water with salt water (35 ppt). When the wind blows, waves roll over your surface. During storms the waves get huge. Things on you are far from land, and you are very deep. It is very dark in your deeper water.

You are _____

2. You have salt water (35 ppt) that rises and falls with the tides. Sometimes the waves roll way up on your sandy shore while at other times much of your sand is not covered with water. Children play on you. When a storm comes, your sand is moved all around.

You are _____

Aquatic Habitat Cards

7. Salt water from the ocean mixes with fresh water from a river in your wide shallow waters. You have lots of food for fish and crabs in your open waters above your muddy bottom. You are a nursery for many ocean animals. Big ships travel to cities over your open water.

You are _____

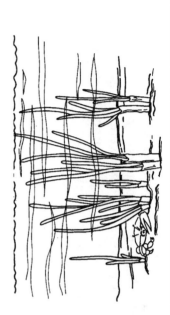

8. Underwater fields of plants grow in your shallow, brackish or salt water on your sandy or muddy bottom. Many animals find food and shelter among the plants. The plants help protect the nearby shore from erosion because they break the force of the waves.

You are _____

5. You have cold, salt water (35 ppt) and a rocky bottom. Your plants and animals are always covered by cold water. You have forests of seaweeds called kelp attached to your bottom. Hundreds of kinds of animals live in, on and among your kelp.

You are _____

6. Your warm, clear, shallow, salt water (35 ppt) and rocky bottom provide the perfect place for animals called corals to grow. Their skeletons provide great habitat for fish. Because you are where it is warm all year around, you are called a tropical habitat. People often swim out from shore to enjoy your colorful fish.

You are _____

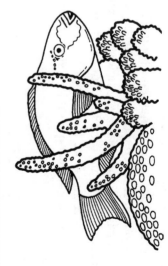

Aquatic Habitat Cards

11. Your fresh water flows over a wide, muddy bottom. Big catfish lurk in your murky waters. Big cities are located on you because you were the easiest place to travel in the past. Ships tow barges up and down you in many places even today.

You are _____

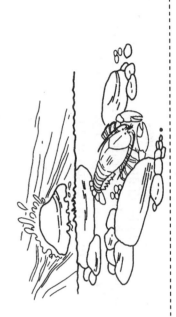

12. Your fresh water tumbles down over rocks and through small pools where fish and crayfish hide. Children wade in you. Your water comes from rain that runs off the land and from springs that bring underground water to the surface.

You are _____

9. Your brackish water is full of nutrients for the tall grasses that emerge along your shore. In the winter these grasses die, but each spring they come back from their strong roots. The decaying grass particles are food for your crabs and oysters. The grasses protect the shore from storms. During very hot dry weather, your water may get saltier than sea water from evaporation.

You are _____

10. Short trees line the shores of your brackish or salt water. Their big roots hold the trees in the mud, even when hurricanes disturb your constant warm days. Many animals and plants find a home on your tree roots or in your waters. Because it is warm all year around, you are said to be a tropical habitat.

You are _____

Aquatic Habitat Cards

13. Your quiet, deep, fresh waters are home to many fish which hide deep beneath your surface. Ships and boats travel over you. Storms may make waves on your wide surface. Where it gets very cold, you may be covered with ice in winter.

You are _____

14. Sun shines through your shallow, open, fresh water, allowing underwater plants to grow across the bottom. Still and small, you may freeze solid where winters are cold. In the summer turtles bask on your shore, deer drink from you and children splash or fish in you.

You are _____

15. Grasses grow out of your still, fresh waters. Red-winged blackbirds build nests in the grasses. The air is filled with the calls of the male blackbirds.

You are _____

16. Tall trees stand in your quiet, dark water with moss at their base. Freshwater turtles bask in a patch of sun while mosquitoes buzz. It is very dark in the shade of the trees.

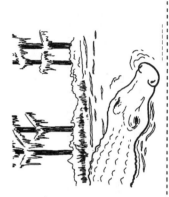

You are _____

Aquatic Habitat Cards

18. Your water is very salty, saltier than the sea. Water may flow into you, but there is no way for it to leave except by evaporation in the hot sun. Salt in the water is left behind. You form in low areas in deserts and behind barrier sandbars which cut you off from the sea.

You are _____

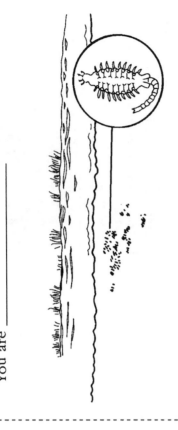

17. Bushes and peat mosses grow in and around your shallow, still water. Patches of very wet ground are home to pitcher plants which get their nutrients from the insects they catch in their leaves. Your water is fresh, but may be very dark and acid.

You are _____

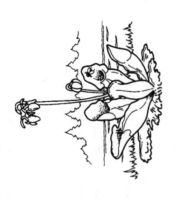

Water Habitats

1. Water is salty (it has salt in it) 2
1. Water is fresh 12
2. Water is salt water (35 ppt; sea water) or saltier than 35 ppt 3
2. Water is brackish, less salty than the salt water or salt water 9
3. Water is saltier than sea water (more than 35 ppt) . . . . . Salt lake or salt pond
3. Water is salt water, 35 ppt 4
4. Open water, not near shore . . . . Ocean or sea
4. Near shore 5
5. Part uncovered at low tide 6
5. Always covered with water 7
6. Sandy bottom . . . . Sandy beach
6. Rocky bottom . . . . Rocky intertidal
7. Bottom of sand or mud . . . . Continental shelf
7. Bottom hard and rocky 8
8. Cold water and cold winters (temperate) . . . . Kelp forest
8. Warm waters and warm climate year-round (tropical) . . . . Coral reef
9. Open water . . . . Estuary
9. Near or at shore with green, rooted plants 10
10. Plants are entirely under the water . . . . Sea grass bed
10. Plants grow out of the water (emerge) 11
11. Climate is cold during winter (temperate) . . . . Salt marsh
11. Climate stays warm all year (tropical) . . . . Mangrove swamp
12. Water flows in a definite bed 13
12. Water appears not to move at all unless windy 14
13. Large, flowing over muddy bottom . . . . River
13. Small, flowing over sandy or rocky bottom . . . . Stream or creek
14. Has open water although shores with plants are around it 15
14. Plants grow out of the water all over 16
15. Large and deep; plants grow under water only near shore . . . . Lake
15. Small and shallow; plants grow under water everywhere . . . . Pond
16. Plants are grasses . . . . Freshwater marsh
16. Peat moss and woody plants (trees or bushes) 17
17. Plants are trees with definite trunk . . . . Swamp
17. Plants are bushes; lots of moss grows on ground . . . . Bog

Note: 35 ppt is a way of expressing how salty the water is in the ocean or other saltwater habitats; if you had one kilogram (1000 grams) of sea water, 35 grams of the weight would be salt.

Aquatic Habitats

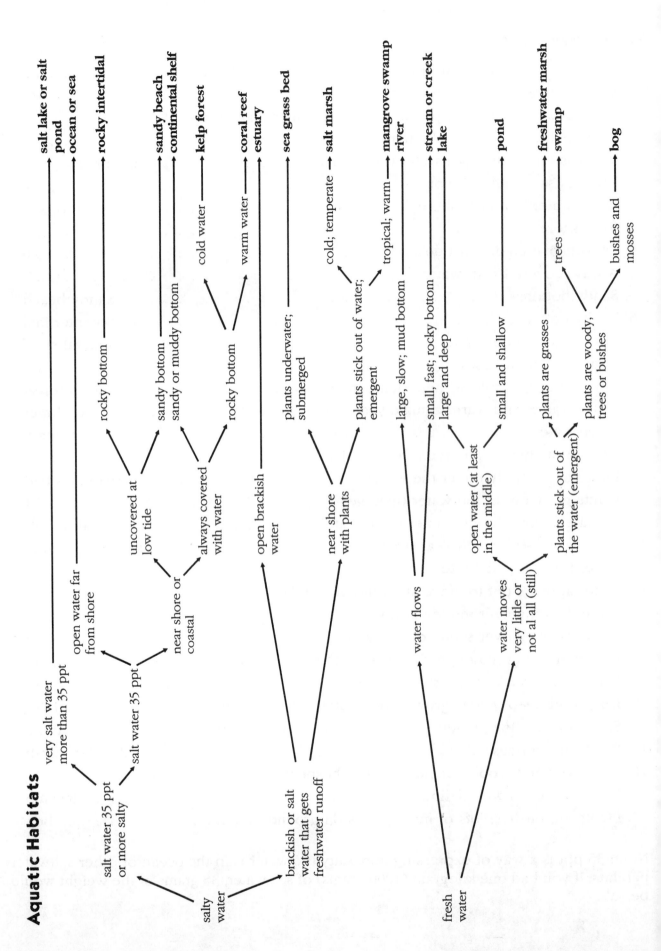

ACTIVITY 3 Your local water habitats

Finding your local aquatic habitats using maps and the library.

Teacher's Information

A common misconception on the part of both students and adults is that there are no aquatic habitats where they live, especially if they live in the heart of a big city or in a desert. After being introduced to aquatic habitats in Activity 2, students should be able to use state and/or local maps to begin to find places where aquatic habitats occur by looking for the physical features they studied in Activity 2 such as a shoreline or a river. Also have them use personal knowledge. For example, ponds too small to show on maps are found in cemeteries and on golf courses. Even a drainage ditch may develop into an aquatic habitat over time, collecting plants and animals drawn to its water.

Introduction

Assign your students to work in groups. Challenge them to see which group can identify the most aquatic habitats within a set geographic limit (i.e., a 50 mile radius from your school). The first problem they must solve is how to tell what areas are within the 50 miles. Can they solve this problem? Make a string cut to measure the fifty miles using the scale on the maps provided and rotate it around the point where your school is. The objective is to list all the kinds of aquatic habitats from Activity 1 that might be in this area. Students may use geographic features on the map to infer the existence of habitats. They may also base their lists on personal observation.

Action

Give the students a set amount of time. Let them work independently during this period, making a list of the habitats they would expect to find. Ten to fifteen minutes should be enough. Have each group write its list on the board.

Science skills
organizing
inferring

Concepts
Aquatic habitats are common everywhere.

Habitats vary with geographic region.

Skills practiced
using maps
library research
using scale on a map

Time
1 class period

Mode of instruction
independent group work

Sample objectives
Students find local or regional aquatic habitats and describe the location of at least one in detail.

Builds on
Activity 2

Materials
for the class
state maps
local maps
rulers
string

Preparation

Collect the maps. Many states have 800 numbers where you can order state maps. Tourism and travel departments for your state, county or city may also help. If you are careful, they will last for years so you only have to collect them once to get a classroom set. Alternately, you may pick the geographic region to be considered based on the social studies curriculum at your grade level. For example, your students may be studying a region other than your city or state in another class.

Outline

before class

1. collect maps

during class

1. practice reading a map and judging scale

2. review water features on map

3. have students define their range and identify aquatic habitats within this range

4. review choices

Results and reflection

Do they agree? Allow each group in turn to challenge another group to defend one of their choices until the class agrees that the lists are reasonable. Allow the class to decide how they could determine the accuracy of the lists. Which group had the longest list with the fewest errors? They win a prize. The nature of the prize is up to you.

Conclusions

Combine all the lists to see how many aquatic habitats are within the region you choose. Can the students agree on which was the most important to the growth of your city or town? Why? Which do they prefer to visit? Why?

Extension and applications

If there is an aquatic habitat within walking distance of your school, organize a hike. On casual inspection have your students do an inventory of numbers of kinds of plants and animals (do not worry about formal identification). Did they correctly identify the habitat based on the key? Can they observe and list human influences on this habitat? Do humans appear to help the plants and animals or hurt them? Make a chart listing what they perceive as good and bad influences. If there are problems, can the students suggest ways to make this a better place? Would they consider adopting this spot for a monitoring program and environmental improvement project?

ACTIVITY 4 Weather watching

How do temperature and daylength change with changing seasons of the year? Long-term weather observations and measurements.

Teacher's information

The physical characteristics of a site determine the kinds of plants and animals that can inhabit that site. Many of these physical characteristics are related to climate and weather. The primary goal of this activity is to track physical data that will be used in Section II and at the end of Section III. Summation of the outcome of this project is Activity 26. You may do this as simply as having students record weekly comparisons of air and water temperatures and daylength. If your school has a complex weather station and/or a pond, this could be a major class project. The point is to track seasonal changes, particularly in temperature, rainfall and light as measured by daylength. This project should run for at least 8–10 weeks.

Introduction

Ask the class to predict what the changes in air and water temperature and in daylength will be during the next three months. Their predictions should depend on the time of year that you start *Living in Water*. How can students test their predictions? They can directly measure temperature and can get daylength newspaper or TV weather reports.

Does the rain itself change with the seasons? Have the students heard of acid rain? How might they measure it?

Science skills

observing
measuring
organizing
predicting

Concepts

The physical characteristics of a place are important in understanding its ecology.

Skills practiced

charting data

using thermometers and other weather equipment

Time

1 class period

10–15 minutes per week over 2–3 months at least

Mode of instruction

independent individual or pairs work

Sample objectives

Students make direct observations of seasonal changes in their local environment.

Students monitor an environmental pollution parameter over a period of time.

Students compare seasonal changes of temperatures in air and in water.

Materials

for class
large bucket
thermometer (see Recipes)
newspaper weather information
rain gauge
pH test kit
goggles
optional
school weather station
pond or stream

Preparation

Determine a location where a large bucket of water can sit outdoors without being dumped or vandalized. It may be that one student in class will have to keep the bucket at home if the school does not have a safe site. Access to a pond on a weekly basis is the best case, but not

Action

Work with the class to establish a plan of action that includes specific responsibilities for who does what when, a chart on which data is recorded, and standardized ways of taking data. Practice how to use the thermometer. Always measure the air temperature before the water so that evaporative cooling does not give you a lower reading in air. Practice the pH test with known solutions like white vinegar and tap water and mixtures of the two.

If you have the time, have each group of students work independently to construct a data chart. Have them exchange charts and critique each others' work. Then discuss the different approaches and come up with a single design. Draw it on a large sheet of construction or butcher paper and post it. Then keep it up to date each week. A data chart might look like this:

week	air temp (°C)	water temp (°C)	daylength (hr)	total rainfall (cm)	pH of rainfall
1					
2					
3					
4					
5					
6					
7					
8					
9					
10					

Day of week measured _____ Time of day recording done _____

Test the rain's pH (acidity) using a wide range test kit when there is sufficient rain to do the test. Rainfall can be recorded in a simple log with the date, amount of rain and pH of the rain. These data will be used with Activity 15.

Results and reflection

This will be completed as Activity 26.

Extension and applications

1. Have students design and build rain gauges. Problems to keep in mind: most rain gauges have a wide top that funnels down into a long, skinny tube so that the scale is amplified, making small amounts of rain easier to read. That means the students are going to have to calculate the area of the top of the funnel and the cross sectional area of the tube. The tube would then be marked accordingly. For example, a funnel that is 5 square inches in area would collect 5 cubic inches of water if 1 inch of rain falls. If that rain then pours into a tube that is 1 square inch in cross sectional area, the rain will take up 5 inches of the tube. Those 5 inches will be marked as ONE inch of rainfall. Challenge students to explain why rain gauges are designed with wider tops for collection and narrow long tubes for reading. Both enhance accuracy. Need a clue? Ask them to accurately measure 10 ml of water in a 100 ml cylinder, then pour it into a 25 ml cylinder and compare accuracy. The expanded scale enhances accuracy.

common. Find the section of your local paper in which sunrise and sunset are listed. Pick one day of the week when your class schedule permits a group of students to spend five minutes recording air and water temperatures and calculating and recording number of hours of daylight. Pick a standard time each day, preferably in the middle of the day. If you have several sections of science, you might compare different times of day. The best thermometer is one designed for field use in water, but the inexpensive ones used elsewhere in this curriculum are fine if used with care. Avoid mercury and unprotected glass.

Pick a site for the rain gauge. It may be a school site or it may be that each student gets to keep it for one week and daily record any rainfall at home. Have the students decide on a consistent spot such as away from house and not under trees.

Outline
before class
1. find a spot for the bucket and fill it

during class
1. discuss student perceptions of weather patterns

2. ask students to develop methods for collecting weather and water temperature data

3. develop and agree on data table

4. practice reading thermometer and using pH test kit (see Recipes)

5. collect data for at least 8–10 weeks for Activities 15 and 26

ACTIVITY 5 Setting up a freshwater classroom aquarium
Creating a model aquatic habitat as a class project.

Science skills
measuring
organizing

Concepts
Scientists build living models of natural ecosystems called microcosms (small) or mesocosms (medium size) to test ideas about how real habitats function.

Studying how living models behave can help with understanding real environments.

Skills practiced
following directions
measuring nitrate
measuring pH
measuring dissolved oxygen
measuring temperature

Time
3 periods over 1 month
daily feeding and record keeping
weekly maintenance

Mode of instruction
teacher-directed class project

Sample objectives
Students practice good record keeping.

Students develop good husbandry practices.

Students design a model of a natural habitat.

Builds on
Activity 2

Teacher's information

One of the best possible motivations for your students to become involved in *Living in Water* activities is an aquarium in your classroom. Many activities have suggestions for using your aquarium for related work. Your students will develop a better understanding of the animals and plants used in *Living in Water* when the creatures are members of their classroom family. Care and feeding of the aquarium becomes a reward for good behavior or a well done assignment and teaches responsibility. Setting up and maintaining a freshwater aquarium is easy. The instructions here are for a freshwater tropical aquarium. If you are an old pro at aquarium keeping, you might choose to try a saltwater system. While you might choose to set up the aquarium without student input and help, your students are going to be your best helpers for taking care of it. Ownership comes with decision making and input from the start.

If you wish to have a special aquarium that models a specific aquatic ecosystem, *The Complete Aquarium* by Peter W. Scott, Dorling Kindersly Publishers, 1995, ISBN 0-7849-0013-8 is a beautiful, inexpensive, outstanding reference. Your students can choose from a variety of geographic regions and compare photographs of real places with their own aquarium designs. The instructions given here are for a mixed species, freshwater tropical aquarium typical of a slow moving river or a lake.

If your budget is limited, ask the students if they have any old aquarium equipment you could borrow. You might be surprised to find some skilled aquarium keepers among your students or their older brothers or sisters. Fish are the most popular pets in most urban areas, and it is not uncommon to find urban high school students running a dozen 100 gallon aquariums in their basements. Also, check the school closets and with other teachers for unused equipment.

Read this entire section before starting, and plan the tasks. You will need to complete the first phase a full month before the aquarium can be completely stocked. Start setting it up with the class helping. Discuss the function of each part as it is added. The students will enjoy participating as their tank takes shape before their eyes.

Introduction

Remind students of some of the aquatic habitats they learned about when they used the habitat cards and pass out the cards to refresh their memories. Would the students like to make a living model of one of these places in the classroom? Which are practical and which are not in terms of all the things that have to be duplicated? For example, can you have waves in your model? List some of the physical characteristics that you might be able to duplicate: fresh water, heat, light. Then either have the students do all the research—a time-consuming project—or use these instructions to build your model.

Action

Phase one

1. Rinse the aquarium tank with water. Do not wash it with soap. Soap is deadly to fish and very hard to remove. If the aquarium is old and needs scrubbing, use a soapless sponge and table salt to scour it. Rinse thoroughly.

2. Pick the location for the aquarium—never in direct sunlight. Indirect light from a north facing window is ideal. Do not put it over a heat register or radiator. Pick a location away from rapid temperature changes. The aquarium will be very heavy, so the location must be sturdy and level. The tank cannot be moved if there is any water in it so make sure you have it where it is going to stay. An electrical outlet must be nearby.

3. Put the tank in place and add the undergravel plastic filter plate and air lift columns according to instructions on their boxes.

4. Put the quartz gravel in a bucket and rinse it with tap water until the water is clear. Pour the water off.

Materials

for a freshwater aquarium

tank (long and wide, not tall and skinny; 20–30 gallons is a good size)

heater(s) (10 watts per gallon; need more than one for tank bigger than 20 gallons)

floating thermometer

light source (indirect sunlight or plant grow light and timer)

undergravel filter (the same size as the tank)

air lift columns

plastic air tubing

air pump (the biggest, quietest one you can afford)

air valve (gang valve)

medium-size (about 1/4 inch) plain quartz gravel; 15 pounds/square foot of bottom

clean rocks, clay flower pots or other structures

plastic screen or cover for top of tank

large plastic bucket(s)

large plastic tubing or a siphon to move water in and out of the tank

net

log book to record feeding and water changes

tropical fish "flake" food

2 fish (1 inch long) suited to chosen water temperature

inhabitants for 20 gallon tropical freshwater tank

6 guppies (swordtails or platys 1–2")

Plecostomus (fish with a scraping mouth)

6 small zebras

1–2 small catfish

10 snails

1 crayfish

2–3 large bunches of *Elodea*

2–4 rooted water plants

optional equipment

nitrate test kit

pH test kit

dissolved oxygen test kit

power filter

algae scraper

Preparation

Gather materials.

Outline

See exercise.

5. Gently add the gravel and spread it over the filter plate. It should be at least one inch deep.

6. If you can, add about one cup of gravel from an established aquarium. This will have the kind of bacteria you want to grow on your gravel. If you cannot do this, you will still have bacteria because they are in the air. They will just take a bit longer to get established.

7. Put a bowl on the gravel so that when you pour water in, it will hit the bowl and not dig a hole in the gravel. Fill the tank part way with water. You may use tap water as you are not going to put any animals in today. In the future, you will let your water sit in a clean bucket for several days to "age." This gets the chlorine out so that it does not hurt the animals. You may also use chlorine remover purchased from a pet store. When doing future water changes, put a heater in the bucket several hours before the water change to warm the new water.

8. Use clean rocks and/or flower pots on their side to make hills and caves in several places. Make hiding places for your crayfish. Do not add corals or shells to a freshwater aquarium as they alter the pH of the water which can hurt your fish.

9. Attach air tubing to the air lift columns and the air pump. The little gang valve will help you regulate the airflow. Plug in the pump. Bubbles should be traveling up the air lift columns. Water is pulled down through the gravel and then up the columns by these bubbles. This traps food and feces in the gravel, filtering them out of the water. Bacteria will take up residence on the surface of the stones and break down organic material like food particles. Bacteria also become established that convert ammonia, the toxic waste product of aquatic animal urine, to nitrate which is less toxic. These bacteria are essential and take about a month to grow. You cannot put all your animals in until this month is up!

10. Add water until the top of the air lift column is covered. Adjust the air flow so it is even.

11. Add the thermometer and heater(s). Make sure the heaters are immersed correctly. Refer to the package for instructions. **Never let the water level fall below the heaters. Unplug them when doing water changes and replug when done.** Plug them in and set them for 76 °F if you are going to have freshwater tropical fish. For temperate freshwater species, 65 °F is about right.

12. Let the aquarium sit for a day and check the temperature. Adjust the heaters if needed.

13. Get the top ready. If it has a light built in, plug it in. You need a top to keep your fish from accidentally jumping out. If you have crayfish, the top must fit very tightly as they are escape artists. If you have a lighted top, the light should be turned on and off on a regular daily cycle to avoid stress to the animals. A simple timer made to turn your home lights on will take care of this. Tropical fish should have 12 hours of light and 12 hours of dark.

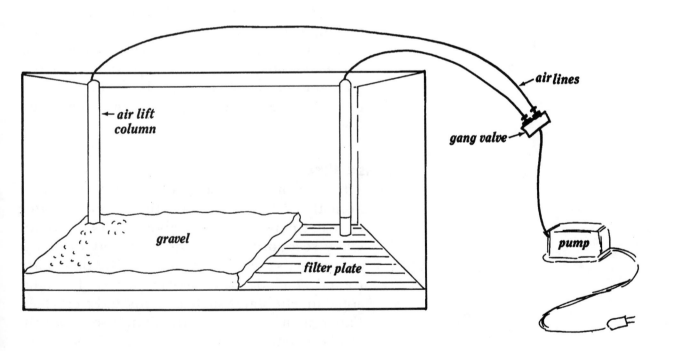

14. Let the tank sit with the air pump running for at least one full day. The water should be clear and clean. If there is still dust from the gravel, siphon the water out and add more.

15. When the temperature is correct, acclimate (see step #19) and add several small (1 inch) fish. Wait at least two weeks to add the final fish. Use goldfish or bait fish for a 65 °F temperate tank or big guppies for a tropical tank. Feed them "flake" food from a pet store. These will be the only occupants for two weeks. If you ignore this two week wait, you may lose all the animals in your tank because of toxic waste build-up.

Phase two

16. After two weeks, your aquarium should begin to have a bacterial population. You can add some more animals now if you do a good partial water change each week and continue to do so. Test for the presence of nitrate if you have a nitrate test kit. If you load the tank up and do not do water changes, the fish will die from the build up of their own wastes. For a water change, drain about 1/3 of the water and replace it with tap water that has aged in a bucket two days before being added. Always check the temperature of the aged tap water before adding. If it is more than 5 °F below that of the tank, immerse the heater in it to warm it. Replace the water slowly with aged tap water to avoid temperature shock. Unplug the heaters before removing water and plug them in again when done.

Phase three

17. After one month, add plants and a few more animals. The volume of the weekly water change can be reduced to about 1/4 of the water every second week. Choose a variety of plants and animals. Some plants float while others are rooted. Include several varieties of fish and some invertebrates such as snails and a crayfish. Experiments in this curriculum use goldfish, snails, crayfish, and plants called *Elodea* (also known as *Anacharis*). Select other species that are non-aggressive, but active swimmers that are fun to watch or have interesting adaptations. Also look for animals that occupy different parts of the aquarium such as bottom-dwelling algae-eaters. If you have designed a habitat with

lots of places to hide and plants that form structure, a diverse community of animals is possible. Do not use species that have been bred by humans for weird anatomy, such as fancy goldfish. Regular goldfish do better at a lower temperature than tropical fish. If you purchase the fish from a pet store, the staff can help you make selections. Most biological supply catalogs will also have some information on each species offered.

If you are collecting local species, make sure the temperature of your tank matches that of the fishes' natural environment so that the animals are not temperature shocked. Taking fish from a cold pond in the late fall into a heated classroom could kill them. Also make sure collecting is legal. Whatever kind of fish you choose, get small ones and avoid species like oscars that grow to dinner plate size. Also avoid large predators. If you bring home a largemouth bass from a farm pond, it will grow to eat all its tankmates. **Resist the temptation to overload the tank.**

18. To add plants, unplug the heaters. Drain about 1/2 of the water from the tank and "plant" the rooted plants in the gravel. Try for natural looking clumps rather than an even spacing. Gently, refill with aged tap water or the drained water, using a siphon. Go slowly if the temperature is different from the tank. Plug the heater back in.

19. To add fish, acclimate the animals to the tank by floating the fish or other animals in plastic dishes or their shipping bags on the surface of the tank. Roll the top of the plastic bags to trap air so they float. Gradually add a bit of tank water until the temperatures are the same. Slowly release the animals into the tank. Acclimation should take about 30 minutes.

Daily tasks

These tasks can be done by students. Initially you should supervise them. They will develop the ability to do them without your help.

1. Feed daily with just a little bit of food. Feed tiny bits at a time for five minutes and then quit. Remove uneaten food. Use commercial flake food and a variety of other things. We know one teacher who kept her tank going entirely with worms and insects the children caught plus carefully selected, non-greasy bits of meat from their lunch sacks. **Do not overfeed.** For example, do not give extra food on Friday. Healthy animals will do fine over the weekend. Unused food will decay, causing bacterial growth.

2. Remove any dead or dying organisms.

3. Check temperature.

4. Record everything done each day in a class aquarium log (notebook). Record water changes, addition of plants and animals, feeding, daily temperature, and any behavioral or health observations about the animals. If you have any problems, the key may be in the notes. This also encourages observation and recording of data, good scientific practices. Staff members at public aquaria practice very detailed record keeping as part of their work.

Weekly/biweekly tasks

1. Do a partial water change. Use a siphon (see Recipes) to "vacuum" the gravel surface, getting the "grunge" off the bottom. Replace the water slowly with tap water that has been sitting out overnight in a bucket. Remember to check temperature. Use a slow siphon to gradually replace the water. Rate of flow can be regulated by tying a knot in the tubing to pinch it shut or using a tubing clamp.

2. If you have a power filter, stir the gravel gently, avoiding disturbing the plants. Put the power filter in place and let it run overnight.

3. If you have test kits, have the students measure and record nitrate, pH and dissolved oxygen each week before doing the water change.

Where to find help

If you have purchased your supplies from a locally owned pet store, the staff will probably be a good source of information. Pet stores and biological supply houses have books on aquarium keeping. Check the card catalog at your local library under aquariums. Ask the staff of a local nature center or math/science center if they keep aquariums. Ask your science supervisor for suggestions about persons who might be knowledgeable about aquariums. Join your local aquarium society. Large public aquariums are a poor source of help because they receive so many calls that they are generally unable to answer them promptly. If you purchase animals and plants from biological supply houses, they may send you information with the shipment, such as Carolina Biological Supply's *Freshwater Aquarium Handbook*.

Sources of equipment

The best way to get an aquarium is to have a family donate it. Many an attic has an old aquarium. Yard sales and want ads are also sources. Make sure that the tank has no cracks in the glass or plastic. If it leaks along a seam, a tube of silicone sealant can cure the problem. Remove the old sealer and use generous amounts of the new to replace it. Is there equipment gathering dust on a school shelf? Check around your school system. Ask your science supervisor. Most teachers we know have found free tanks within their system, leaving their limited funds to purchase the rest of the supplies. Aquarium equipment can be purchased at pet stores. Most scientific supply houses also offer aquarium supplies. Both offer packages with all the parts in one order. Most package deals will need an additional heater to be adequate for cold school rooms over a long weekend. Packages are easiest, but more expensive.

Sources of plants and animals

Pet stores and biological supply catalogues carry a wide range of animals and plants suitable for aquariums. Stick to the cheap, common species unless you are going to get very involved in this project. Stores that sell live bait for fishing are a cheap source of an interesting variety of animals that are tough and inexpensive.

If you live at the seashore or in the country where there are farm ponds, you may collect animals yourself. Make sure you are collecting legally. Federal, state and county parks generally prohibit collecting without a permit. Many states require permits or fishing licenses regardless of location. Freshwater pond animals are great aquarium specimens if you get small ones. Marine species from cold waters or that live along rocky shores where there are waves are hard to keep alive without special tanks. Animals from warm estuaries, such as killifish, or that live completely below the low tide line in tropical areas are more likely to survive.

During school vacations

In these energy-conscious days schools can become very cold over the weekend or a break. Regardless of what kind of fish you keep, you should purchase a heater(s). Buy more watts than needed: 10 watts per gallon for tanks over 20 gallons. Heaters are designed for homes, not freezing schools, so make sure you have power. There is not much you can do about schools that turn off the air conditioning on the weekends.

Fish and invertebrates do not use food energy to keep warm, so they eat less than birds or mammals. One of the most frequent mistakes made by new aquarium owners is overfeeding. Fish and invertebrates can do fine without eating over the weekend. For a week break, feed and do a water change just before you leave and immediately on returning.

At the end of the school year

Tropical species may be adopted by students or their parents for home aquariums. Pet stores may accept the return of healthy specimens. If the animals came from the wild, they may be returned only if they go back to exactly the same place you got them and are slowly acclimated to the water. Never release any plants or animals that are not native to the exact body of water they are entering. Introduction of foreign species can do incalculable ecological damage.

Results and reflection

Students may do library research on the kind of environment, such as the Amazon River, that the aquarium represents. Each student might also write a paragraph comparing their aquarium to the real place. Then have the students discuss the kinds of habitat they have created. How is the aquarium like the real thing? How is the aquarium different? Can the students make suggestions for how the model (aquarium) could be more like the real place it models? Have them evaluate which of the suggestions are realistic. What would they have to do to implement them? Support changes that are realistic and modify your tank.

Conclusions

Models are a good way to begin to understand how real things might work, but they are seldom perfect enough to completely replace studying the real thing.

Using your classroom aquarium

Consider animal adaptations. How do the animals that were selected use the space in the aquarium? Have the students chart the places in the tank and which species use which places. Put a plastic sheet on the side of the tank and mark it with different colored water soluble markers for each species at 1 minute intervals for 15 minutes to show how the space is used. Are there unused spots that might accommodate another species? What kind of adaptations would it need to use that space? Are there conflicts among species in the tank for sites? How could that be changed?

Extension and applications

1. For a different kind of writing experience, one teacher suggests reading the part of T.H. White's *Once and Future King* where Merlin teaches Arthur several things by turning him into a fish. After hearing this story, have the students write about what is would be like to be a fish in your classroom aquarium.

2. Have students contact your nearest public aquarium by email through the Internet or regular mail and ask for career information on being an aquarist, the person who cares for the fish and invertebrates. Ask for a job description and what kinds of courses and experience are required. What opportunities does the aquarium offer for students to become involved in classes or as volunteers? Most are not able to take students until they reach high school, but now is a good time to find out what is available. Some have special programs targeting middle school students.

3. Do other classes have aquaria? How about starting an aquarium club?

SECTION II Things dissolve in water

Teacher's information

Solutions

Many substances DISSOLVE (go into SO-LUTION) in water. Solutions are generally transparent, appearing clear, although the chemicals in solution may color them. Solids such as salts or sugar, liquids such as alcohol or acetone, and gases from the air all dissolve in water. Water is frequently referred to as the "universal solvent" because so many different kinds of things dissolve in it. Some things do not dissolve in water. The saying that "oil and water do not mix" is based on fact. A mixture that has as much of a compound in solution as is physically possible is SATURATED with that particular compound. The amount of one thing in solution can influence the amount of other things in solution. For example, salt water cannot hold as much dissolved oxygen as fresh water which has no salt in solution.

Waves, wind, and currents may actively mix things in solution. If there is no mixing, compounds spread slowly through water by DIFFUSION, which is mixing caused by the movement of molecules. Since diffusion is 300,000 times slower in water than in air, it takes quite a while for a chemical to reach equal distribution throughout a body of water without mixing.

Substances in solution in a pond, stream, lake or ocean have a direct effect on the plants and animals that live there. Each water environment is unique, in part, because of these substances. This section discusses common things in solution in water, such as air or salts, and the ways that these substances affect plants and animals.

Salts

SALTS are chemical compounds that go into solution in water easily. They consist of IONS: charged atoms or molecules. Positive ions are missing one or more electrons. An example is the sodium ion, written Na^+. Negative ions have one or more extra electrons. An example of a salt is sodium chloride or $NaCl$, also known as table salt. There are many different kinds of salts besides table salt. When salts are added to water, they dissolve very rapidly because the ions are not firmly bound to each other and are attracted to the water molecules.

Water cycle

Rain or melting snow picks up salts and other compounds because they easily dissolve in water as it moves over the soil and enters streams as RUNOFF. Water that sinks into the ground (PERCOLATES) also picks up chemicals as it moves down and becomes part of the GROUNDWATER. Rain even dissolves chemicals in the air as it falls. Hence, air pollution can lead to water pollution such as acid rain.

Eventually these weak solutions of salts and water reach the ocean or a lake such as the Great Salt Lake that has no outlet. Liquid water can turn to a gas, WATER VAPOR, when exposed to air. When water turns into water vapor (EVAPORATES into the air), it leaves the salts behind. The water vapor rises into the upper atmosphere. In the colder upper air this water vapor CONDENSES back into a liquid (rain) or freezes into snow. This PRECIPITATION (rain, snow or sleet) falls again on the land where it dissolves up more salts. Each time as the water evaporates, the salts remain behind. This water cycle constantly carries more salts to the sea along with other chemicals. Over millions of years the oceans have become quite salty.

Ocean water

Ocean water is referred to as SALT WATER because of this saltiness. If you evaporate 1000 grams of ocean water, 35 grams of salts and other chemicals remain. This means that 35/1000 of the weight of sea water is salts and other chemicals. The percentage of salts by weight is calculated by dividing 1000 into 35 (0.035) and then multiplying by 100. Thus salt water is 3.5% salts and other chemicals by weight. Saltiness (SALINITY) is also expressed in parts per thousand so salt water is also said to be 35 parts per thousand (ppt). The saltiness of ocean water is reasonably constant at 35 ppt over the earth's surface though oceanographers measure minor variations among water masses. In enclosed lagoons and bays where evaporation is high, it may be saltier. For example, the water in Flor-

ida Bay (between the Everglades and the Florida Keys) may reach 60 ppt during a summer drought when freshwater runoff is low. Salt lakes may be much saltier than the sea.

Water very low in dissolved materials is called FRESH WATER. Where a river (fresh water) runs into the sea, the saltiness is lower than salt water's 35 ppt. It is called BRACKISH WATER. The area of brackish water at the river mouth is an ESTUARY.

Table salt, NaCl, does not account for all ocean saltiness. If you try to keep sea creatures in a solution of table salt, they will die because they require the whole range of kinds of salts in the sea. There are many kinds of things in solution in salt water. The major ions found dissolved in sea water are shown in the table below.

positive ions		negative ions	
kind	concentration ppt	kind	concentration ppt
sodium (Na^+)	10.6	chloride (Cl^-)	17.3
magnesium (Mg^{+2})	1.3	sulfate (SO_4^{-2})	2.7
calcium (Ca^{+2})	0.4	bicarbonate (HCO_3^-)	0.7
potassium (K^+)	0.4		

Fresh water

While fresh water has little dissolved salts and other chemicals, it is not absolutely pure. Even rain picks up things from the air as it falls. Different bodies of fresh water have different compositions of things in solution, depending on local rock and soil, plant material entering the water and the human activities that influence the system. By definition fresh water is less than 0.5 ppt dissolved salts and other chemicals. This may not seem like much, but fresh water that has a good deal of calcium leaves a white residue behind in tea kettles and coffee makers and requires extra soap in the washing machine. It is often called "hard" water. By comparison "soft" water has little calcium. The differences from one freshwater system to another can directly influence the kinds of plants and animals found there. These differences also mean that one system may be less susceptible to acidification than another because some chemicals can counteract (BUFFER) the acid, neutralizing it.

Dissolved gases

Atmospheric gases are in solution in water in the same proportions as in air (nitrogen is highest, followed by oxygen, etc.) When atmosphere and water are in contact for long periods and no other things are affecting the system, the gases reach an equilibrium and saturate the solution. The saturation values are temperature dependent as well as dependent on other factors like salinity. When gases enter water under pressure, the possible amount of gas in solution is much higher than at normal pressure. If the pressure is released, the dissolved gases may take quite a while to come out of solution. If there is more gas in water than normal for a certain pressure, a SUPERSATURATED solution exists.

Nitrogen

Water falling from great heights over dams or through dam turbines becomes saturated with gases under great pressure. The pressure decreases below the dam and the water is supersaturated as it moves downstream, an unstable condition. In this unstable condition gas bubbles may form and rise explosively. Think of removing the lid on a warm carbonated beverage. In the case of fish that absorb these gases under pressure, nitrogen forms bubbles and causes the bends just as it does with SCUBA divers that surface too rapidly. Water supersaturated with nitrogen kills fish below many high dams in the western United States.

Oxygen

Oxygen is an important gas for things that live in water. In surface waters oxygen is present in the same proportions as in air, but under many circumstances the amount of oxygen varies in different parts of a body of water. Biological activity, incomplete circulation and the slow DIFFUSION or movement of gases through water all contribute to this unequal distribution of oxygen. For example, in a shallow biologically rich body of water, the oxygen level may be higher than expected. During the day plants produce much more oxygen during PHOTOSYNTHESIS than is used by both plants and animals in RESPIRATION. At night dissolved oxygen falls rapidly as the plants and animals use it in respiration. In oceans, bays and lakes, areas may form in which oxygen is low because mixing and diffusion do not replace it as fast as it is used.

A problem common in many bodies of water is pollution that results in excess oxygen use or low DISSOLVED OXYGEN (DO). In this case oxygen is used faster than it diffuses into the water. It is used by DECOMPOSING organisms which are acting upon pollutants such as sewage or animal wastes. The decomposing organisms may also be feeding on the remains of plants that grew in over-abundance due to excessive water fertilization by plant nutrients from human sewage or farm fertilizer. Environments that are very low in oxygen are said to be ANOXIC and are difficult places for most animals to survive. Bodies of water with very low dissolved oxygen may experience major fish kills, especially in summer. Warm water holds less oxygen and at the same time fosters faster decomposition.

pH

Each water MOLECULE is made up of two ATOMS of hydrogen and one of oxygen: H_2O. Water is highly stable, but very small amounts "come apart" under normal circumstances forming two kinds of ions: H^+ and OH^-. Acidity is the concentration of H^+ ions in solution and is expressed as pH (a proportion of H^+). Pure water has a pH of 7 and has 10^{-7} H^+ ions and an equal concentration of OH^- ions. Pure water is NEUTRAL because of this equality. A solution of strong ACID would have lots of H^+ ions (e.g., a pH of 1 would be 10^{-1} H^+) and very few OH^- ions. Conversely, the solutions we call BASES are low in H^+ ions (e.g., 10^{-13} H^+ ions is a pH of 13) and high in OH^- ions.

This system of pH numbers is confusing because the pH number is based on a power of 10, that is, each incremental number is 10 times different from the next. For example, a pH of 4 is ten times as acidic as 5. This means that pH is not an intuitive concept, especially for children. It is also hard to get used to the idea that the smaller the number, the stronger the acid. For example a pH of 1 is 1,000 times as acidic as a pH of 4. Because both strong acids and strong bases react with living tissues, destroying them, pH is important in aquatic systems.

As rain falls through the air, gases (such as carbon dioxide, sulfur dioxide and nitrous oxides) dissolve in the rain. This changes the pH of the rain. Naturally occurring carbon dioxide gas in air acidifies rain to a pH of 5.6. Normal rain should have a pH of 5.6. Chemicals from air pollution caused by burning coal or oil high in sulfur (sulfur dioxide) or from any high temperature combustion which oxidizes some of the nitrogen gas in the air that passes through the furnace (nitrous oxides) can dissolve in rain, forming acids. Rain with a pH as acidic as 4.3 occurs in parts of the United States. In areas where soil or rock does not naturally neutralize the acidity, aquatic organisms can suffer.

Acidification also occurs where rocks with certain chemical composition react with rain when it reaches the ground. Usually, this happens where the rocks have been disturbed by mining. Acid mine drainage may be a serious problem in some locations. Surface water with a pH of 6.5 show some damage to aquatic organisms. Below 6, many species suffer. Acidification is not a problem in salt water because sea water has chemicals which neutralize acids in the carbonate buffer system.

Mineral nutrients

Many other ions and elements also dissolve in water, including those that are essential for plant growth. Two MINERAL NUTRIENTS are the most essential for plant growth in aquatic systems: PHOSPHORUS as phosphate (PO_4^{-3}) and NITROGEN as ammonium (NH^{+4}) or nitrate (NO_3^{-2}). If these are present in low amounts, the growth of algae and green plants may be limited, reducing the amount of food produced in the system. If they are present in high amounts, overproduction of algae or of undesirable species may disrupt an aquatic system. In general, phosphate is the limiting nutrient in freshwater ecosystems and nitrogen is limiting in salt water with brackish systems varying, depending on salinity.

Water pollution

In addition to acid rain and nutrient enrichment, all sorts of other things end up in solution in water. Many of them are natural products, although human actions may cause levels of these that are higher than would occur naturally, so we regard them as pollution. Mercury and nitrates are examples. Others are products which we have manufactured that are entirely new such as pesticides and herbicides and other organic chemicals.

Water pollution starts on land. Runoff may carry farm chemicals into streams and rivers. Factories and cities dump their sewage into SURFACE WATERS (rivers, bays and oceans). Water that percolates down into the soil to the GROUNDWATER can carry chemicals from farms or waste disposal sites. Pollution of groundwater is a major problem. Over half the people in the United States get their drinking water from groundwater wells.

Water pollution may constitute a direct health hazard to humans as well as the plants and animals that live in water environments. It may kill living things directly, but it is more likely to have effects that are less immediate. Pollution can cause diseases or it can change reproduction. It may also disrupt the relationships between living things, changing the community structure.

One way of classifying pollution is to separate materials that come from an identifiable source (POINT SOURCE POLLUTION) from those that enter water all along its course (NON-POINT SOURCE POLLUTION). In the past most attention was focused on point source pollution because it was easier to see and identify. Regulators were able to write laws that specified how much pollution came out of a factory or sewage treatment plant. As point source pollution is beginning to come under some degree of control, attention is turning to non-point source pollution that enters aquatic systems in runoff and groundwater. It is important, but hard to define and regulate.

There are hundreds of kinds of water pollution. This curriculum concentrates on three: nitrate enrichment and its consequences; conditions leading to low dissolved oxygen and the impact on aquatic animals; and acidification and the ways it affects both plants and animals in aquatic systems. These three can be easily investigated by children and have broad application to many areas. They can be tested with safe kits and are subject to investigation in both the classroom and the field.

ACTIVITY 6 Salty or fresh?

Which is heavier (has greater mass), fresh water or salt water?
Student experiments which test the relationship of salinity and
density of water.

Science skills

observing
measuring
inferring
predicting
experimenting
communicating

Concepts

If temperatures are the same, an
equal volume of salt water is heavier
than fresh water.

Fresh water floats on salt water
because it is lighter (less dense).

Skills practiced

use of spring scale, simple balance or
triple beam balance
measuring volume

Time

1 class period

Teacher's information

This exercise examines the relationship of salinity to
weight (density) of solutions. Subsequent studies
show how these relationships determine the distribution of fresh and salt water where a river enters
the ocean and forms an ESTUARY. Students get
practice designing tests and comparing outcomes.
They also learn about selecting the appropriate
equipment for a job and ways to increase their
accuracy when conducting experiments. (The term
MASS is more correct than weight but is not used.
You may substitute it if you wish. Since only mass
on Earth is discussed here, weight seems to be
sufficient.)

Introduction

Ask your students how they would determine which
was heavier: salt water or fresh water? They will
probably suggest weighing the solutions. Use a balance and two clear plastic cups of water, one with
twice as much water as the other to demonstrate. Ask
them if this is a "fair test." Why not? The students

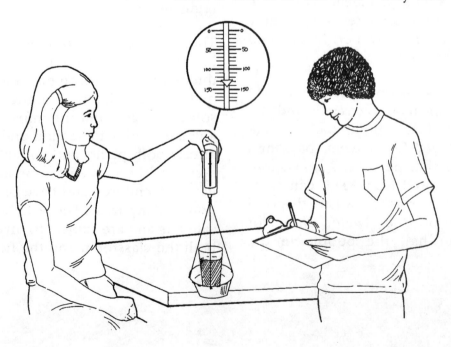

should be able to tell you that you must use the same amount or volume of water to tell which is heavier. Then demonstrate with a plastic glass and a heavy ceramic coffee cup filled with carefully measured equal volumes of water. Is this a "fair test"? Again, they should say no and identify the difference in weight of the two containers as the problem. What rule would they have to follow for this to be a "fair test"? There can be only one difference between the two items tested. In this case, the difference is the amount of salt in the water.

Action

Can they design a "fair test" to answer the question you first asked: which is heavier, salt water or fresh water? Let individual groups try without telling them exactly what to do. Let them make mistakes and discuss these mistakes among themselves. They may compare equal volumes of salt and fresh water on a simple balance or they may weigh equal volumes with a scale or triple beam balance.

After most groups have answered the question, have everyone settle down and state his/her results. Salt water is heavier. If there are disagreements, have students explain their methods and carefully repeat work. What should become apparent is that the larger the volume they compared, the more accurate their measurement. Can they tell you why? Ask them to speculate. The margin of error in a small sample is amplified by the small size. For example, a 3 ml error on a 25 ml sample is more than 10% off while a 3 ml error on a 200 ml sample is only 1.5% off. The 100 ml cylinder was a better tool than the 25 ml. Best results are achieved by using about 200 ml water in a 9 ounce cup. Also, they should recognize that the graduated cylinders are much more accurate than the kitchen measuring cups. A basic lesson from this exercise is that the tools you choose to use are important to the quality of work you do.

Now can the students predict what would happen when fresh water meets salt water? This happens where a river flows into the sea. Have them state their predictions. How can they test their predictions using clear plastic glasses, salt and fresh water, food coloring and a teaspoon? Give them these materials and

Mode of instruction
teacher directed group work

Sample objectives
Students state experimental evidence proving that salt water is heavier than fresh water.

Students control variables.

Students compare relative weights.

Materials
for each group
4 stick-on labels
marking pen
small bottle of food coloring
4 clear plastic cups (6 or 9 oz)
plastic teaspoon
2 cup measuring cup
25 ml plastic graduated cylinder
100 ml plastic graduated cylinder
simple balance
250 gm Ohaus spring scale
hanging pan (made from 3 strings about 2 ft long and an individual serving size aluminum pie pan)
for each two groups
1/2 gal salt water (1 cup salt per gallon) at room temperature in plastic milk jug
1/2 gal tap water (fresh water) at room temperature in plastic milk jug
for class
heavy coffee mug
2 clear plastic cups

Preparation

Mix the salt water. Groups share jugs so two gallons of each is enough for the class. If you wish to substitute triple beam balances for the simple balances described here, train students to use them ahead of time. Simple balances that compare two items at once may require some testing by students before they believe that the heavier item pulls its side down and lifts the other up. Simple spring scales may require a bit of practice with reading the scale. Ohaus 250 gram spring scales are best. If the spring scales lack pans, make them from small aluminum pans (individual pot pie size) and three pieces of string 2 ft long. Tie string equidistant around the rim through three punched holes. Tie in a loop at the top of the strings with two overhand knots. The loop must be above another knot so that all three strings are pulled equally.

let them experiment. Do not tell them what to do right away.

They may add food coloring to the salt water as a marker. Most will try to pour one into the other. This may not give clear results. For a clearer outcome, see what happens when a little salt water is carefully laid on top of a cup of fresh water with the teaspoon. (See the illustration for technique.) The spoon is lowered into the water and rotated out from under the salt water. The salt water should sink to the bottom, indicating that it is heavier.

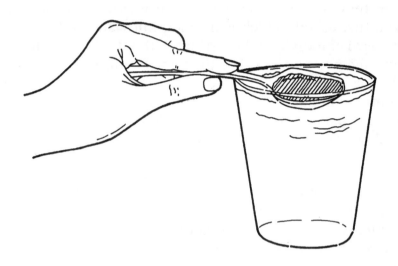

Can they be sure the colored water sank to the bottom because it had salt in it? What if it was the food coloring that made it heavier? How could they prove it was the salt? Have groups meet with each other to discuss this and design a new experiment. Try the reverse with colored fresh water and plain salt water to prove that the food coloring is not causing the result.

Results and reflection

Have your students tell you what they discovered after they have had the opportunity to fill out their discussion sheets. Generally, they should be able to state the following and give evidence from their tests to prove it. If equal volumes of salt and fresh water are compared in containers that are identical, the salt water weighs more than the fresh water. When salt water is gently placed on the surface of fresh water, it sinks. Fresh water gently placed on the surface of salt water floats at the surface.

Conclusions

Salt water is heavier than fresh water when both solutions are the same temperature. (Temperature as a variable will come in a later activity.) When two things of equal volume are compared, the heavier is said to be more DENSE.

Using your classroom aquarium

Is your classroom aquarium fresh water or salt water? Have your students suggest tests to find out. It could be compared with fresh and salt water in the same tests done above.

Extension and applications

1. If your students got an accurate measure of the weight of fresh and salt water, ask them to calculate the weights of different volumes of water. How much does a liter of salt water or fresh water weigh? For a tougher problem, how much does a gallon of each weigh? A liter of fresh water at 0 $^\circ$C should weigh 1000 gm or 1 kg.

Outline

before class

1. copy worksheets

2. make balance pans if needed

3. mix salt solution and let all water sit at room temperature

during class

1. show the students the equipment and ask if they could design tests to discover which is heavier (has greater mass): salt or fresh water

2. use unequal volumes and unequal containers to spark discussion about fair tests (controlling variables)

3. have students work in groups to answer posed question

4. have students predict what happens when salty and fresh water meet; test predictions with layering

5. review results from different groups and resolve differences

6. draw conclusions and how students addressed variables

Activity 6
Salty or fresh?

1. Make a table showing what volume of each kind of water you weighed and how much it weighed. Be sure to include units.

2. Using weighing, which did you find heavier: salt water or fresh water?

3. If the groups did not agree, describe what additional tests you did to reach agreement.

4. Use drawings to describe what happened when salt water and fresh water were very gently layered on each other? Draw and label what you observed.

5. Using layering, which was heavier: salt water or fresh water?

6. Explain what you did to prove it was the salt and not the food coloring that gave the results you observed.

7. Predict what might happen when the fresh water from a river meets the salt water of the sea.

Activity 6
Salty or fresh?

Name **possible answers**

1. Make a table showing what volume of each kind of water you weighed and how much it weighed. Be sure to include units.

	salt water	fresh water
volume	150 ml	150 ml
weight	160 gms	152 gms

2. Using weighing, which did you find heavier: salt water or fresh water?

salt water

3. If the groups did not agree, describe what additional tests you did to reach agreement.

We did not agree. Different groups weighed different amounts. At the end we all decided to try to weigh lots of water. 200 ml fit into the cups so that is what we used. Most groups got salt water heavier. On the balance it was heavier, too.

4. Use drawings to describe what happened when salt water and fresh water were very gently layered on each other? Draw and label what you observed.

 ← fresh water
← salt water

 fresh water
salt water

5. Using layering, which was heavier: salt water or fresh water?

Salt water was always heavier with layering.

6. Explain what you did to prove it was the salt and not the food coloring that gave the results you observed.

We did two tests. One had coloring in the salt water. In the other, it was in fresh water. Salt was always under fresh.

7. Predict what might happen when the fresh water from a river meets the salt water of the sea.

The salty ocean would have fresh water from the river layered on top of it the way it did in the glass.

ACTIVITY 7 The layered look
A teacher demonstration of the distribution of fresh and salt water in areas where the two meet in estuaries.

Science skills
observing
predicting
communicating

Concepts
Stratification occurs in estuaries where fresh water meets salt water.

Fresh water flows above the saltwater layer below it.

Some mixing occurs where the two layers meet.

Time
1/2 class period (do with Activity 8)

Mode of instruction
teacher demonstration

Sample objectives
Students describe and explain the distribution of salt and fresh water in an estuary.

Builds on
Activity 1
Activity 2
Activity 6

Teacher's information
An ESTUARY is defined as a semi-enclosed body of water where incoming sea water is diluted with fresh water from the adjacent land. Because of the differences in weight (density) between fresh and salt water, salt water moves upstream in the estuary along the bottom, while fresh water flows downstream along the surface. This causes a layered condition. Some mixing occurs at the interface where fresh and salt water meet. The mixed water flows out with the fresh water along the surface. This layered condition is said to be STRATIFIED. This teacher-led demonstration illustrates the stratification that may occur in estuaries. You can go directly from this activity to Activity 8 as this is the set-up for the next demonstration.

This activity again provides an opportunity to discuss the need for controlling variables when doing tests.

Introduction
Start by asking the students what happens when salt water meets fresh water. They may remember from Activity 1 that salt water and fresh water mixed give BRACKISH water. After Activity 6, they may also suggest that the heavier salt water will settle near the bottom, while the fresh water floats on top of the salt water.

Let's see what happens when a large body of salt water meets fresh water in a model ESTUARY. Have the students place themselves so that they can see without bumping the demonstration table. Describe what you are going to do and ask them to predict the outcome.

Demonstration
Fill one container 1/3 full with clear, aged fresh water. Then slowly siphon in the colored salt water solution, keeping the siphon tube near, but not on,

the bottom of the container, holding it against the side. Do not move the siphon, and do not let it flow too fast. The rate at which the water in the siphon flows is determined by the height of the intake end. Keep it only slightly above the end where the water is flowing out. A colored salt solution layer should form on the bottom of the container. Observe the container from the side. Have the students describe what happens. Did this match their prediction? Is this a fair test? Discuss. The answer is "no" because one has food coloring as well as having salt. The food coloring is a second variable.

How can you solve this problem of two variables? Have the students propose solutions. Someone should come up with doing the reverse test with clear salt water and colored fresh water in another container. Do it. Does this prove the conclusion from the first experiment? You should end up with two stratified systems with the color on top in one and on the bottom in the other. The best way to see this is to look from the side, not the top.

Materials
for the class
2 clear plastic containers such as large deli containers or small aquaria; may use clear glass 1 gallon jars

2 siphons—clear plastic aquarium tubing

1/2 gal of clear aged fresh water

1/2 gal of aged fresh water with 4 drops of green or blue food coloring

1/2 gal of clear aged salt water (1 cup salt per gallon)

1/2 gal of aged salt water (1 cup salt per gallon) with 4 drops of green or blue food coloring

optional
hose clamp

for each student
colored pencils, markers or crayons
worksheet

Results and reflection

Ask the students to observe the containers of water. Questions to get a discussion going might include: How many layers formed? Two. Which layer is salty? Bottom. Which layer is fresh? Top. Are they completely separate? No. Is something happening at the interface between the two layers? Yes. What? Mixing is occurring. How can you tell there is mixing? If this is a good model of an estuary, what would happen if one measured the salinity at differing depths from

Preparation

Make up the saltwater solutions and label bottles. Because this exercise works best as a teacher-led demonstration, plan a spot where all the students have a clear view on a solid surface that they cannot move as they crowd around. Aged water which has been sitting open at least overnight insures that the next activity will not hurt the animals. Practice your siphoning technique. Suck water up into the tube until it almost reaches your mouth. Quickly drop the end you have been sucking below the level of the intake end and put it into the receiving body of water. Remember that the source needs to be elevated above the container into which you are siphoning but not too high above or you will cause turbulence and mixing from fast flow out the siphon. You can slow the flow with a loose knot in the tubing. Or, better yet, use the hose clamps made for tubing to adjust the flow. Also, it is critical that the solutions all be the same temperature.

surface to bottom in an estuary? Salinity would increase with depth.

Why did we do this twice? The food coloring added a second variable. When it was used in the reverse order in the second experiment, you proved it was not the cause of the results.

Have the students record the results on their data sheets, using crayons or markers to indicate the location of the colored water layer. Have them label each layer of water in each container, and then fill out the rest of the data sheet.

Conclusions

Because salty water is more dense than fresh water, it sinks below fresh water when the two come into contact. While there is some mixing at the boundary, a stratified system with regard to salinity is formed. The distribution of salinities in an estuary reflects this relationship.

Extension and applications

1. Challenge the students to decide how they could more clearly see the mixing. Try using two different food colors in the fresh and salt water solutions which when combined produce a third where mixing takes place (for example, yellow and blue to make green).

2. How long will this stratification last? Put the system with two colors aside for several days. You might be surprised at the results.

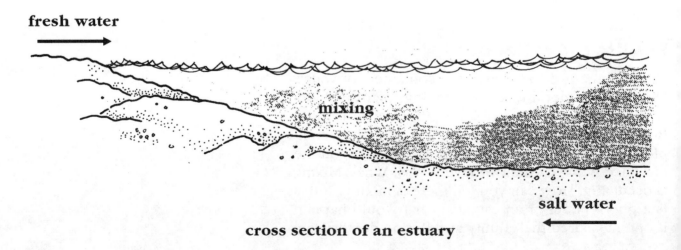

fresh water

mixing

salt water

cross section of an estuary

National Aquarium in Baltimore

3. What might cause mixing in an estuary? Ask the students to think about what might stir things up. The most common mixing agent is wind. Have students try blowing across the surface of your stratified system. You might also try a fan, but don't pick it up while turned on.

4. Draw the distribution of salt and fresh water on the board in a cross section of a typical estuary as shown here. Include the directions of water flow. Have the students consider how the animals in an estuary might use these currents. Can they think of what animals could do with these currents of water in an estuary? Travel or MIGRATE. Animals that cannot actively swim, like ZOOPLANKTON, but that can go up and down in the water, could travel up and down the estuary by staying near the top or the bottom.

Outline
before class
1. mix solutions
2. copy worksheet
during class
1. reinforce the term ESTUARY and ask the students to predict what happens when fresh and salt water meet in an estuary; remind them of Activity 6

2. demonstrate by siphoning two layered systems: colored fresh water on plain salt water and plain fresh water on colored salt water

3. discuss why it was done both ways for a fair test

4. complete worksheet and do Activity 8

Activity 7
The layered look

1. State the question you are trying to answer by observing this demonstration.

2. Draw the results of both demonstrations here and label the parts:

    first container   second container

3. Based on the results of this demonstration, where would you expect to find the saltiest water if you were studying the point at which a river joins the ocean and forms an estuary? Explain your reasoning.

Activity 7
The layered look

1. State the question you are trying to answer by observing this demonstration.

    The salt and fresh water are added to each other gently. What happens when you do this slowly? This would be like a river meeting the ocean in a bay.

2. Draw the results of both demonstrations here and label the parts:

    first container   second container

3. Based on the results of this demonstration, where would you expect to find the saltiest water if you were studying the point at which a river joins the ocean and forms an estuary? Explain your reasoning.

    I think the salty water would be like the salt water in our demonstration — on the bottom. Every time we have tested it salt water has been heavier and heavy stuff sinks.

ACTIVITY 8 Some like it salty—some do not!

A teacher demonstration showing how salinity may affect the distribution of some aquatic animals within estuaries.

Science skills

observing
inferring
predicting
experimenting
communicating

Concepts

Salinity is one factor which helps determine where different animals and plants live in an estuary.

Some animals can sense different saltiness and move to their preferred location.

Time

1/2 class period (do following Activity 7)

Mode of instruction

teacher demonstration

Sample objectives

Students explain how adaptation to a particular salinity may affect where an aquatic animal lives.

Builds on

Activity 1
Activity 2
Activity 6
Activity 7

Teacher's information

Salinity is an important factor in determining the distribution of living things in an estuary. Marine organisms inhabit the high salinity region at the mouth of the estuary. Traveling up the estuary along a decreasing salinity gradient, the number of marine species decreases. A few marine organisms, such as blue crabs, that can TOLERATE different ranges of salinity are very abundant throughout the estuary. Freshwater species are found in tributaries leading into the estuary and near the surface estuarine water. In part, due to this wide range of salinity, there are many different habitats in an estuary. At one location, marine species may be more common at the bottom while freshwater species are at the surface because of salinity stratification.

While some species of animals and plants tolerate wide ranges of salinity, others have narrow ranges of requirements. Animals seek salinities within their range of tolerance. In the case of plants and animals like oysters which are attached to the bottom as adults, their location is determined by salinity present when the seeds germinated or the animal larvae settled. Animals that can swim may move to remain in the OPTIMAL or best salinity. Other factors that influence habitat selection include temperature, light, food supply, predators and oxygen levels.

Introduction

Show the goldfish and adult brine shrimp to your students. Challenge them to predict where brine shrimp and goldfish might choose to be in each of the two stratified systems. Ask them to give their reasoning for their answers. The amount of salt in the water may influence where the animals stay in the container. The students have clues to the animals' preferences: "brine" shrimp and the kind of water the students use for their goldfish at home.

Strain the adult brine shrimp, pouring the water through the net and drop half in each container. Observe their behavior for several minutes. Ask questions about their choice of location.

Using the net, gently transfer one goldfish into each container and let the students observe their swimming patterns. The students must remain quiet and not scare the fish. Questions to get a discussion going might include: How many times does each animal swim to the bottom of the container? Where does it end up swimming most of the time? Also, do they think the brine shrimp affect the fishes' behavior? What is their evidence? The goldfish may try to eat the brine shrimp, indicating that food also influences distribution.

Result and reflection

Have students record their observations. Ask them in which water habitats might they look for these fish and brine shrimp in nature. Pass around the habitat cards for ponds and salt lakes. How did their salinity differ? Goldfish prefer fresh water while brine shrimp live in salt lakes and can tolerate very high salinity. Why did you use two tanks, one with colored fresh water and the other with colored salt water? To avoid having two variables. Color might affect distribution

If you have very hungry goldfish, they may dive down onto the salt water to eat brine shrimp and then pop back up to wash the salt out of their gills in fresh water. Look for the washing action, rapidly puffing water in and out. They pop up out of the salt water because they are very buoyant in strong salt water.

Conclusions

Aquatic animals that can swim choose the place where they live based on many factors, one of which may be salinity. In a stratified system some species might always be at the surface while others are always at the bottom. Even though they are at the same spot on a two dimensional map, in three dimensions they may be vertically separated and not come in contact with each other. When possible, aquatic animals will avoid water in which the salinity is outside their range of TOLERANCE and will seek OPTIMAL situations.

Materials

for the class
both demonstration jars or tanks from Activity 7
small dip net
2 goldfish 1 inch long (do **not** substitute other species)
several dozen adult brine shrimp
aquatic habitat cards (Activity 2)
for each student
worksheet

Preparation

This demonstration uses the two stratified systems made in Activity 7. It shows that animals may select a position in the water based on the distribution of salinity. Do this as a teacher demonstration and **do not** substitute species. Goldfish tolerate salt and will not be harmed by short exposure. Other fish species may not survive this. Purchase both the fish and the brine shrimp at a pet store. If you cannot find adult brine shrimp, skip that part of the activity.

Outline

before class

1. save set-up from Activity 7 or remake it

2. copy worksheets

during class

1. show the students the brine shrimp and the goldfish and discuss habitat preference

2. ask students to predict which part of the layered systems each kind of animal would likely prefer

3. put some of the brine shrimp and one goldfish in each jar or tank and observe behavior

4. remove the fish and brine shrimp by straining through the net

5. complete worksheets and discuss, using habitat cards for review

Using your classroom aquarium

The fish from this exercise may come from your aquarium and may return there when done. The brine shrimp will make welcome food for your tank's inhabitants.

Extension and applications

1. What are the effects of too much salt or too little salt on an organism? Your students can observe this by immersing small weighed cubes of potato in very salty water (the water from this experiment) and in fresh water. Generally, plant and animal tissue loses water (weight) in salt water and gains it in fresh. The causes of this are somewhat difficult to adequately discuss at the upper elementary/middle school level and require an understanding of the diffusion of water through membranes. The "what" is easy to see, but the "why" is not. You might also put leaves from an aquatic plant in both fresh and salt water if you have microscopes to study the results. Draw the before and after of each. How might these observations relate to the goldfish's choice of water?

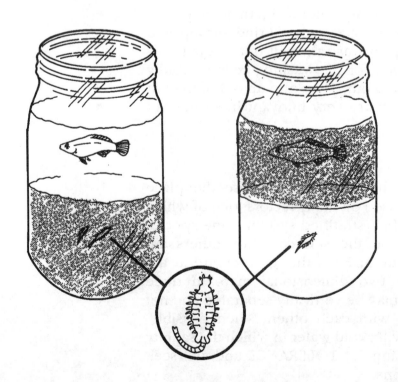

Activity 8
Some like it salty—some do not!

Name _____

1. Predict which layer(s) the animals will prefer and give your reasoning.

   brine shrimp

   goldfish

2. Draw the two tanks. Label the salty and fresh water. Draw the animals in the locations they preferred.

3. Did the animals always stay in the same kind of water? If not, describe their behavior and give a reason for why they did not.

Activity 8
Some like it salty—some do not!

Name __possible answers__

1. Predict which layer(s) the animals will prefer and give your reasoning.

   brine shrimp

   I don't know what these are, but the teacher said some kids call them sea monkeys so I think they would like salt water because the sea is salty.

   goldfish

   Goldfish live in fresh water. I have 2 at home.

2. Draw the two tanks. Label the salty and fresh water. Draw the animals in the locations they preferred.

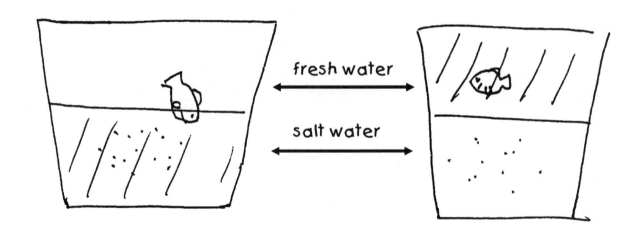

fresh water

salt water

3. Did the animals always stay in the same kind of water? If not, describe their behavior and give a reason for why they did not.

   The goldfish stayed on top in the fresh water most of the time except when they dived down into the salt water to eat the brine shrimp. And then they popped back up to the fresh water and washed the salt out of their gills.

National Aquarium in Baltimore

ACTIVITY 9 The great salinity contest!

A contest in which each student applies his/her understanding of salinity, using experimental techniques to find the winners.

Teacher's information

This activity is an application of student learning during previous activities. Now that students have learned about the relationships of salinity and density, have them put their knowledge to work in a contest by discovering who has the saltiest and the freshest water samples. Students receive the samples at random, so the activity is a lottery combined with a logical puzzle. The prizes to the winners may be objects or rewards of special opportunities.

Introduction

Hand out the student sheets and tell them they have a mystery to solve. The idea for this mystery came from what Hurricane Hugo did to samples at a marine laboratory in South Carolina. They are to work on their own and together, following the directions. There are several things they cannot do: they cannot taste their samples! Never taste unknown solutions. Also, getting a fresh water sample from the sink is not legal. You may place the samples on student desks (and be accused of favoritism) or let the students come up and select their numbered samples.

Action

Students follow directions on worksheets.

Science skills

observing
measuring
inferring
predicting
experimenting

Concepts

Water containing dissolved salts is heavier (more dense) than fresh water.

Density increases with increased salinity.

Skills practiced

measuring volume
weighing
using a balance
reading for information

Time

1 period

Mode of instruction

independent individual student work

Sample objectives

Students apply their knowledge about weight (density) of salt and fresh water to determine the relative salinity of unknown solutions.

Students plan their testing procedure before employing it.

Builds on

Activity 1
Activity 6
Activity 7

Materials

for each student

numbered sample (about 1.5 c or 400 ml) in a large plastic cup (about 16 oz)

2 small clear plastic cups

plastic spoon

for each group

25 ml plastic cylinder

100 ml plastic cylinder

2 cup measuring cup

250 gm Ohaus spring scale

balance

food coloring in dropper bottles

for the class

2 qt fresh water at room temperature

2 or 3 gal salty water at room temperature (1 cup kosher or canning salt per gallon; see Recipes)

1 qt very salty water (1/2 cup kosher or canning salt per quart; see Recipes)

prizes such as stickers, a puzzle or maze, pencils, free time

Preparation

Duplicate the worksheets. Collect and sequentially number (1, 2, 3, 4, etc.) one large cup for each student. Prepare the solutions in gallon and half gallon milk jugs, being careful to get them exactly the same for the intermediate sample. You might mix the two or three gallons of the intermediate samples in a bucket to insure uniform salinity. This test is sensitive enough to measure small differences. The volumes above are sufficient for 21 intermediate samples, 1 very salty and 6 freshwater samples. Reduce the sample size or mix up more intermediate salinity for a larger class. Use kosher or canning salt to

Results and reflection

How are they going to find out who won? They can do any physical test which will get at the relationship of density or heaviness to salinity: weighing samples, layering on each other and comparing equal volumes on a balance. When they are finished, they can check with you to find out if their results are accurate since you recorded which solution was in which sample bottle. Award the prizes when they have correctly found the fresh water and the saltiest water.

Conclusions

Planning before starting was critical to success. Have a brief class discussion of techniques used for testing. Which was best? Which was fastest? How important was planning? Did they have enough sample? How was planning important to having enough? Follow different lines of reasoning. Which is more important: getting the right answer or understanding how to approach the problem? What do they think?

Extension and applications

1. How about asking the question: which is saltier, the Dead Sea (276 gm of salt per kg of water) or the Great Salt Lake (266 gm salt per kg of water)? This can be solved in a number of ways. Everyone could compete to see who can dig the answer out of the library fastest. Or you could mix up two samples representing each using the information above and have the students test them. Locate each site on the map.

2. How do salt lakes form? Water flowing over the ground dissolves salts which are carried to a low spot from which there is no way for water to flow out. As the water evaporates, the salt is left behind.

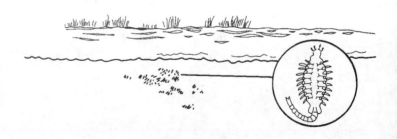

3. Do salt lakes always stay the same? No! Some years the Great Salt Lake has had more water flowing in than has evaporated out. In 1986 the lake was rising fast. To prevent homes, roads and businesses from disappearing under water, Utah considered pumping water out of the valley and into another desert valley. The Salton Sea in California just keeps shrinking. Check both out in reference books.

4. Can the students think of another habitat where salinity gets high? Tide pools can get very salty on a hot day during low tide! So do the pools of water in a salt marsh during low tide on a summer day. Salt marshes or ponds where there is restricted flow from the ocean and no rain or runoff of fresh water can also get saltier than the ocean. They are referred to as hypersaline. Florida Bay, between the Everglades and the Florida Keys, also becomes hypersaline in the summer when evaporation is high because the freshwater input from the land was diverted years ago in order to drain wetlands for farming.

avoid the cloudiness that comes with table salt for salt shakers. Do this far enough ahead of time for all the solutions to reach room temperature. Fill some (4–8) cups with fresh water. Most get the intermediate salt solution. Fill one with the very salty water. Record which cups got which solutions as you fill them so you know which wins. The saltiest wins the grand prize. The fresh water wins second prizes. The rest of the students lost in the lottery, but can "win" by figuring out what they have.

Outline
before class
1. prepare solutions using only kosher or canning salt

2. distribute them among numbered containers with careful records; one gets very salty and 4–6 get fresh water and the rest get the intermediate salinity

3. copy worksheets

during class
1. distribute unknown solutions and worksheets

2. students follow the worksheet instructions

3. no tasting and no water from the tap

4. check results as the students complete their work and arbitrate debates

5. review conclusions and distribute prizes

Activity 9
The great salinity contest

Name _____

A. Read this without talking to anyone:

You are a scientist who has experienced disaster. A hurricane hit your marine lab. All the samples you collected from your last three research trips have been soaked as they blew off into the forest behind the lab. The labels you pasted on the jars came off when they got wet. You have walked through the forest behind the lab and picked up as many samples in plastic jars as you can find. One is in the numbered cup or bottle on your desk. The other people who went with you have also got the same problem with their samples.

One of the research trips was to an African lake that had formed in an extinct volcano in the jungle. The second trip was to an estuary in Mexico where there was a big city and many farms that grew produce for export. On your last trip you all collected samples from a healthy South Pacific coral reef. You wanted to stop any bacterial growth so you added a small amount of a toxic compound to preserve the samples. **You may not taste your sample.**

The hurricane has also destroyed most of your laboratory equipment. Can you figure out which place your sample came from with the little bit of equipment that is left? **You may not use tap water** as the hurricane knocked out your water supply.

1. Your sample number is _____.

2. Working alone:

    If you do not work with anyone, what can you do with the equipment provided to identify your sample? Write a possible plan of action:

2. Wait until everyone in your group has answered this question. Then do what you planned above.

3. Based on your results, can you identify the source of your water? Was it the lake, the estuary or the ocean? Explain your reasoning.

4. Are you sure that you are right? Explain.

B. Working with only one other person, quietly discuss and answer this section on each of your worksheets:

1. If you work together, what can you do now that you could not do before? How will this help you determine which sample you have?

2. Test your plan. What happened? Did the results contradict or support your first idea? Now which sample do you think you have: lake, estuary or ocean?

C. Working first with anyone in your group and then with anyone in the room, test your sample as extensively as you wish.

1. Before starting, anticipate and describe any problems you must avoid. Describe anything new you try.

2. My sample was from _____

3. My evidence for this was:

Activity 9
The great salinity contest

Name <u>**possible answers**</u>

A. Read this without talking to anyone:

You are a scientist who has experienced disaster. A hurricane hit your marine lab. All the samples you collected from your last three research trips have been soaked as they blew off into the forest behind the lab. The labels you pasted on the jars came off when they got wet. You have walked through the forest behind the lab and picked up as many samples in plastic jars as you can find. One is in the numbered cup or bottle on your desk. The other people who went with you have also got the same problem with their samples.

One of the research trips was to an African lake that had formed in an extinct volcano in the jungle. The second trip was to an estuary in Mexico where there was a big city and many farms that grew produce for export. On your last trip you all collected samples from a healthy South Pacific coral reef. You wanted to stop any bacterial growth so you added a small amount of a toxic compound to preserve the samples. **You may not taste your sample.**

The hurricane has also destroyed most of your laboratory equipment. Can you figure out which place your sample came from with the little bit of equipment that is left? **You may not use tap water** as the hurricane knocked out your water supply.

1. Your sample number is ___17___.

2. Working alone:

    If you do not work with anyone, what can you do with the equipment provided to identify your sample? Write a possible plan of action:

    > If I am very careful measuring and weighing my sample and I use a big sample, I might be able to tell if I have fresh water (which we found was 1 gm/ml) or water saltier than fresh. But I need to compare to know how salty.

2. Wait until everyone in your group has answered this question. Then do what you planned above.

3. Based on your results, can you identify the source of your water? Was it the lake, the estuary or the ocean? Explain your reasoning.

    > The ocean or the estuary. My water weighed 210 gms for 200 ml so I think it is salty but I don't know if it is brackish or salt water because I need some other salt water to compare to.

4. Are you sure that you are right? Explain.

No. Weighing was not all that accurate last time we did it.

B. Working with only one other person, quietly discuss and answer this section on each of your worksheets:

1. If you work together, what can you do now that you could not do before? How will this help you determine which sample you have?

We can compare weights, we can use a balance to compare and we can layer our water with food coloring for identification.

2. Test your plan. What happened? Did the results contradict or support your first idea? Now which sample do you think you have: lake, estuary or ocean?

I still think I have the ocean or the estuary. My partner had water that was lighter than mine — it could be lake and I have estuary or it could be estuary and I have ocean.

C. Working first with anyone in your group and then with anyone in the room, test your sample as extensively as you wish.

1. Before starting, anticipate and describe any problems you must avoid. Describe anything new you try.

I might run out of sample or get it mixed up

2. My sample was from **estuary**

3. My evidence for this was:

It floated on some people's water and sank on others'. And I found most people had water that was just the same as mine.

ACTIVITY 10 Fresh from salt

Where does rain come from and other water cycle wonders: a model of the water cycle built and tested by students.

Science skills

observing
organizing
inferring
communicating

Concepts

Water exists in nature in all three states of matter: gas (water vapor), liquid and solid (ice).

Water from the ocean evaporates, moves in the atmosphere and falls on the land, only to return to the ocean in a process called the water cycle.

Skills practiced

building a model
reading for information
following directions

Time

2–3 class periods

Mode of instruction

independent group work

Sample objectives

Students work cooperatively to build and test a model of the water cycle.

Students apply personal experiences to an understanding of the states of water.

Students apply a test from a previous exercise to prove the function of the model.

Builds on

Activity 2
Activity 4
Activity 6

Teacher's information

One of the most common demonstrations in a science curriculum is the water cycle. Teachers boil water and condense steam on a pan full of ice cubes to illustrate the concepts involved. This model may be hard for students to relate to the real world. As the ways in which we teach shift more toward student-directed, hands-on, cooperative learning and our testing evolves toward performance assessment, we might want to change the way we teach the water cycle. Here is one possibility. Copy the instructions for your students. Let them work independently in groups. This will take more than one science period. Do not direct student work. Make them figure it out for themselves. Set the model up one day and "run" it until the same time the next day. It could easily take three science periods to complete this. You should have a class discussion to reach closure even if this is being used as a test.

Water occurs naturally in all three states in which matter can exist: gas, liquid and solid. Water circulates between the oceans or freshwater systems, the atmosphere and the earth or living things in the sense that a single molecule over a period of time may move from one to the next. This circulation is named the water cycle.

Introduction

Challenge the students to perform an entire investigation without your guidance. Hand out the worksheets to the groups, show them where the materials are and stand back! One caution—if you are using an artificial light source to represent the sun, remember that hot bulbs can burn both hands and plastic. Remind students to be careful.

Action

Students follow written directions.

Results and reflection

Discuss the worksheet answers completely. Have the students draw a water cycle on the board and label the parts. Have them give examples of personal observations of the three states of matter in which water occurs. Have them relate the parts of the model to the parts of the water cycle. Have them be as detailed as possible about how their model is like and unlike the real water cycle. Let them explain the logic they used in proving the water collected in the cup was fresh water. How could they prove it did not come from melted ice leaking from the inverted soda bottle top? Use the layering test for fresh and salt water from Activity 6 by layering the water caught in the cup on salt water. Where do they have salt water? In the bottom of the model. Discuss how "weather" relates to the water cycle. Have students clean the equipment and store for reuse.

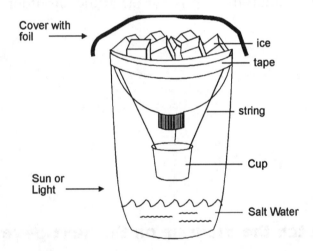

Cover with foil → / ice / tape / string / Cup / Sun or Light → / Salt Water

Conclusion

Ask students to relate why understanding the water cycle is important to them. All of their drinking water comes from fresh water. If water supplies were not renewed by the water cycle, fresh water would run downhill to rivers and out to the sea with nothing to replace it. What drives this cycle? Energy from the Sun.

Extension and applications

1. Bring weather maps from the newspaper or printed from the Internet to class. Ask the students to explain how these are also models.

Materials

for each group

2 liter clear plastic soda bottle—remove label, cut top off at the shoulder; and screw the cap on firmly

1 m of string

0.5 m of regular aluminum foil

0.5 m masking tape

small (2–3 oz) plastic cup

500 ml ice cubes (2 cups)

container for ice (i.e., cottage cheese tub)

300–400 ml salt water (1.5 cups)

container for salt water such as measuring cup

for each student

student worksheet

for the class

a sunny window or a warm light source

Preparation

Cut the tops off of the soda bottles yourself. Punch a hole and then use scissors to cut the plastic. Test a spare to make sure you are cutting at a point that makes the top big enough to hold ice and that will not fall into the bottle when the top is turned upside down. Assemble the materials and mix the solution of salt water. If you are using an incandescent light, set it up in a safe spot.

Outline

before class

1. cut soda bottles

2. copy worksheets

during class

1. provide materials and worksheets

2. let students work independently, following worksheet instructions

3. discuss results and test for fresh water

Activity 10
Fresh from salt

Words that are vocabulary words that may be new are written like THIS.

A. Your group is going to build and test a MODEL.

Scientists make small models of large systems so that they can test their understanding of how the big systems might work. As young scientists, you are going to build a model that lets you test how the oceans, the Sun, and the ATMOSPHERE (the air) interact to produce rain, which is FRESH WATER and has no salt, from the ocean's SALT WATER. This rain falls on land and forms streams and rivers that flow down hill and back to the sea.

For each question, each person should decide how he or she would answer the question, working alone. Then discuss your ideas and form a group answer. Decide who is going to write your answers. Everyone will help decide what they are. Choose a materials supply person to get the following items:

> 2 liter clear plastic soda bottle without a label; the top is cut off at the shoulder and the cap firmly screwed on
>
> 1 m of string
>
> 0.5 m of regular aluminum foil
>
> 0.5 m masking tape
>
> small (2–3 oz) plastic cup
>
> about 500 ml container of ice
>
> about 300–400 ml salt water

B. Assemble your materials to match the diagram on the next page.

Wrap the small plastic cup in foil to protect it from heat and carefully suspend it below the cap. It should not touch the ice-filled top. When the cup is in the right position, use the tape to seal the top edge. Put ice in the upside down top. Cover the ice on top with foil to protect it from heat also. Put your model in a warm spot with direct sun for at least four hours. (On a gray day, you may use a light bulb to play the part of the Sun. Do not put it too close to the plastic or it will melt the bottle.) The model may be left over night.

C. Answer these questions after setting up your model:

1. Your model contains the following parts:

    - a heat source which represents the Sun
    - the salt water which represents the ocean
    - the air (atmosphere)
    - the upper atmosphere (air) is very cold
    - rain gage to measure rain

On this diagram of the model label the parts that represent: ocean, cold upper atmosphere, and rain gauge.

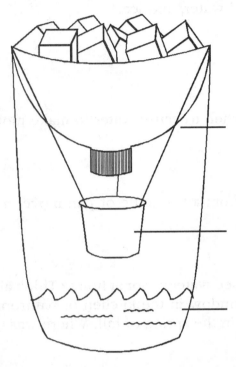

D. Now read this paragraph as a group and as a group answer the questions that relate to each number based on your personal observations and experiences:

The Sun sends heat as well as light to Earth (1). This heat warms the land, water and air (atmosphere) (2). Water exists as a gas (water vapor), a liquid (water), or a solid (ice) (3). Heated water makes more water vapor (4). As the Sun warms the ocean, water EVAPORATES (turns to water vapor from liquid) and becomes a part of the air. The warm air rises (5). Water vapor from the air returns to liquid form (it CONDENSES) when it cools (6). This condensation occurs on dust particles in the air and forms rain drops which fall because they become heavy as they grow.

1. Give evidence based on personal observation that the sunlight warms things.

2. State a relationship between the length of time the Sun shines each day and the warmth of the day based on personal evidence.

3. Have you personally observed the relationship between temperature and the states of water in your kitchen? Describe the temperature conditions under which you have seen water vapor (steam), liquid water, and ice.

4. What would you do in your kitchen to cause water to make more steam?

5. What direction does the steam from a tea kettle or pan move in?

6. Have you ever observed condensed water vapor at home? Think about: your bathroom mirror after a hot shower, the window in the kitchen or bathroom in the winter, the car windows on a cool morning in the spring or fall. Where was the water vapor, and what did it look like?

E. After at least four hours or on the following day, get your group together and examine your model.

Scoop out the ice and ice water. Take the model apart very carefully to preserve whatever might be in the cup. What has happened? Can you relate your observations about the model to the paragraph you read above? Discuss this in your group.

1. What is in the small cup? Can you think of ways to test whether it is fresh water or salt water? Write your plan for testing the water in the cup. Then do the test. You may **not** taste it.

2. If the cup were not in the model, what would have happened to the water that condensed on the cold cap, the rain?

3. The word "cycle" refers to things that can be connected in a pathway to form a circle. The three terms: Ocean, Water Vapor and Rain can be connected with arrows to make a cycle. The arrows can be labeled: falls, condense and evaporate. Write the three terms and connect them with arrows labeled with the name of the process to show how the water cycle worked in your bottle.

4. What part of the world is missing in your bottle model? How could you add it?

5. Can you design a different test for your model? What question would you ask and what would you do to answer it?

6. How did working with the model help you understand the water cycle?

F. Working independently, each of you should now write a paragraph on how the model shows that the ocean, the Sun, and the atmosphere (the air) interact to produce rain, which is fresh water without salt, from the ocean's salt water.

Activity 10
Fresh from salt

On this diagram of the model label the parts that represent: ocean, cold upper atmosphere, and rain gauge.

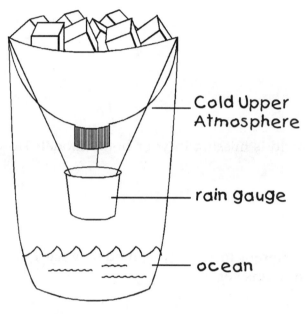

Cold Upper Atmosphere

rain gauge

ocean

D. Now read this paragraph as a group and as a group answer the questions that relate to each number based on your personal observations and experiences:

The Sun sends heat as well as light to Earth (1). This heat warms the land, water and air (atmosphere) (2). Water exists as a gas (water vapor), a liquid (water), or a solid (ice) (3). Heated water makes more water vapor (4). As the Sun warms the ocean, water EVAPORATES (turns to water vapor from liquid) and becomes a part of the air. The warm air rises (5). Water vapor from the air returns to liquid form (it CONDENSES) when it cools (6). This condensation occurs on dust particles in the air and forms rain drops which fall because they become heavy as they grow.

1. Give evidence based on personal observation that the sunlight warms things.

> If I sit in the sun on a cold day, the side of me facing the sun gets warmer.

2. State a relationship between the length of time the Sun shines each day and the warmth of the day based on personal evidence.

> In the summer when it is hot, my mom lets me play outside in the yard until 9 pm because it is light. I have to be inside by 5 pm in January when it is cold. So lots of light means warmer weather.

3. Have you personally observed the relationship between temperature and the states of water in your kitchen? Describe the temperature conditions under which you have seen water vapor (steam), liquid water, and ice.

Ice is in the freezer where it is cold. Water can be either hot or cold and be liquid from the tap in the sink. Steam comes from a boiling pot of water.

4. What would you do in your kitchen to cause water to make more steam?

Turn up the heat on the stove.

5. What direction does the steam from a tea kettle or pan move in?

It goes up.

6. Have you ever observed condensed water vapor at home? Think about: your bathroom mirror after a hot shower, the window in the kitchen or bathroom in the winter, the car windows on a cool morning in the spring or fall. Where was the water vapor, and what did it look like?

The toilet tank drips onto the floor if you flush it lots on hot, humid days. The cold water in the tank causes this.

E. After at least four hours or on the following day, get your group together and examine your model.

Scoop out the ice and ice water. Take the model apart very carefully to preserve whatever might be in the cup. What has happened? Can you relate your observations about the model to the paragraph you read above? Discuss this in your group.

1. What is in the small cup? Can you think of ways to test whether it is fresh water or salt water? Write your plan for testing the water in the cup. Then do the test. You may **not** taste it.

There is a tiny bit of water — about 1/2 inch deep. We are going to layer it on salt water. We can use the salt water from our model.

2. If the cup were not in the model, what would have happened to the water that condensed on the cold cap, the rain?

It would have fallen back into the "ocean."

3. The word "cycle" refers to things that can be connected in a pathway to form a circle. The three terms: Ocean, Water Vapor and Rain can be connected with arrows to make a cycle. The arrows can be labeled: falls, condense and evaporate. Write the three terms and connect them with arrows labeled with the name of the process to show how the water cycle worked in your bottle.

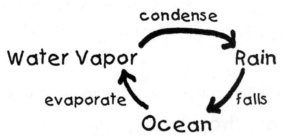

4. What part of the world is missing in your bottle model? How could you add it?

We didn't have any land. We needed an island. A rock would work.

5. Can you design a different test for your model? What question would you ask and what would you do to answer it?

We wondered if we put blue food coloring in the ocean, would the water in the cup be blue or would it stay in the ocean like the salt.

6. How did working with the model help you understand the water cycle?

I could prove that water from the ocean could evaporate and then form rain.

F. Working independently.

Each of you should now write a paragraph on how the model shows that the ocean, the Sun, and the atmosphere (the air) interact to produce rain, which is fresh water without salt, from the ocean's salt water.

The sun warmed our model and helped make water evaporate from the ocean even though I could not see this happen. I could see the water collect on the cold lid and drip down to the cup like rain. It was fresh water because we tested it. It was amazing that the salt all stayed in the ocean, but we would not have any water to drink if it didn't.

ACTIVITY 11 It's raining, it's pouring

Students make watershed models to demonstrate what happens when rain falls on permeable and impermeable surfaces, modeling groundwater and runoff in watersheds.

Teacher's information

Rain or melted snow can do one of two things: sink into the soil or run off the land downhill until it reaches a body of water. The water that sinks in (INFILTRATES) becomes GROUNDWATER which returns to the surface as springs which flow into streams or lakes. Springs release water year-round, providing a constant source of water for wildlife and humans. Early settlers built their homes next to springs. Today groundwater is pumped to the surface for humans from wells. One half of the United States population depends on groundwater for drinking water. Soil or other surfaces that allow water to sink in are called PERMEABLE surfaces. Soil that soaks up water and then releases it slowly as springs acts like a sponge, providing a constant supply of water and reducing flooding.

The water that flows across the land as RUNOFF promptly reaches streams, lakes or bays. Surfaces that produce runoff are referred to as IMPERMEABLE. The more impermeable surfaces in an area, the greater the chance of flooding during hard rain or a fast spring thaw. Asphalt, concrete and roofs are totally impermeable. Soils have a range of permeability based on soil type and the amount of plants growing there. Clay is least permeable, and sandy soils are most permeable. The greater the vegetation, the more rain sinks in rather than runs off. Bare soil and grass catch less water than shrubs or trees. The way humans modify landscapes when they build towns and cities or farms has the unintentional outcome of reducing groundwater and increasing runoff.

All of the land that produces runoff into a specific body of water is named after the body of water. For example: the Chesapeake Bay WATERSHED is all the land from which water drains into the bay. It includes six states and the District of Columbia. This huge watershed consists of many subunits, each named after the stream or river into which it drains such as the Potomac River watershed or the James River watershed.

Science skills

observing
inferring
communicating

Concepts

Water that falls on impermeable surfaces runs off into small streams which lead to ever larger streams and rivers.

Water that falls on permeable surfaces sinks into the ground and becomes a part of the groundwater which may flow naturally to the surface or be pumped to the surface for human use.

A watershed is all the land that drains into a specific body of water.

Skills practiced

construction of models
reading for information
following directions

Time

2 class periods

Mode of instruction

teacher directed group work

Sample objectives

Students make and test watershed models.

Students predict what will happen to rain in their own neighborhoods.

Students propose modifications to their neighborhoods that increase the amount of water that sinks in and decrease runoff.

Builds on

Activity 2
Activity 10

Materials

for each group

disposable aluminum turkey roasting pan or plastic cat litter box

3–5 small plastic glasses

30 inch sheet heavy duty aluminum foil

1/2 gal milk jug sprinkler (see Recipes)

clear plastic 1/2 qt or 1 qt shallow deli container (kind you get slaw or potato salad in)

2 cups aquarium pea gravel (rounded pebbles)

pump from hand lotion or soft soap bottle

aquatic habitat cards from Activity 2

for each student

worksheet

Preparation

Teach this as an alternation between discussion led by you and group work. Make the sprinklers if this is the first time you are doing this activity (see Recipes).

The idea for the aluminum foil watershed came from *The River Times* by the Richmond (VA) Math and Science. They found it in a 1950's U.S. Forest Service environmental education curriculum during an ERIC search.

Outline

before class

1. make sprinklers

2. copy worksheet

during class

1. have the students follow the worksheet; stop and discuss at the end of section A and B

2. discuss the usefulness of models

Introduction

Ask your students to write a paragraph about what happens to rain when it falls near their home. This can be done as a homework assignment or in class. Have the students keep these. Give them time to discuss and write answers on the worksheet at each step in the exercise. Save the introduction of the formal vocabulary words until the final discussion.

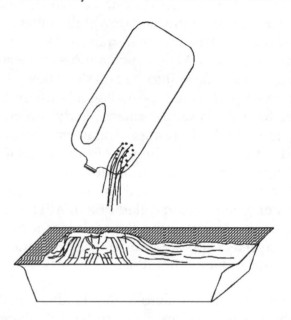

Action

Challenge the students to work in groups to make models of what happens to rain. First, ask them to use the foil, turkey pan and cups to make a model of what happens to rain when it falls on hard surfaces. Use the cups turned upside-down for mountains or hills in one end of the pan. Put the foil over the cups and mold it into land forms such as valleys, rivers, lakes and a bay at one end. Use the habitat cards from Activity 2 to review the different kinds of aquatic habitats they wish to make in their model. Have them think about their local landscape and name some of the parts. For example: in California the mountains might be the Sierra Nevada with the Sacramento River running into the San Francisco Bay. When their landscapes are made, use the sprinkler to make it rain. Have them make sure the lid is on tight, do not hold fingers over the vent and tip it straight up so the water does not run down the side. Groups should discuss their observations and then each student can record his/her observations.

Next, have students make a model of what happens when water falls on a surface which allows the water to sink in. Each group can make a groundwater model by holding a pump which has been removed from a hand lotion or soft soap bottle in a clear deli container and then pouring gravel around it. Make it rain with the sprinkler. Do this inside the turkey roasting pan to catch the spills. Where did the water go? Discuss and record observations. Then how do humans get this water? They pump it. Let the students pump the water up into a cup.

Results and reflection

Students have made observations of runoff from hard surfaces and infiltration of water on permeable surfaces in the models. Now they should reflect on and discuss the following:

- What parts of the models used here represent what parts of the real world?
- How are the models like and unlike the real world?
- Why is it useful to make models in this case?
- Can you suggest another way to model where rain goes?

Have the students clean the foil, bowls, gravel, and trays and store for use in the next activity. The heavy duty foil lasts for several years of use.

Conclusions

Lead students in a discussion of their observations and introduce the formal vocabulary: WATERSHED, RUNOFF, PERMEABLE, IMPERMEABLE and GROUND-WATER. Have students read through their written assignment about how water behaves where they live. Can they use any of these words in place of words they wrote? Have them write about and justify one change they could make around their home or in their neighborhood that might increase the amount of water going to groundwater and decrease the runoff.

Extension and applications

1. An excellent class project would be to design and then implement a plan to increase the infiltration of rain or snow melt and decrease

runoff on the school grounds. Your students could start the plan with this lesson and improve it throughout the term. They might also undertake the actual plan. Many states have environmental education funds which allow schools to apply for money to make positive changes in the local environment. (Check with your state or regional environmental education supervisor.) Some of the items in the plan don't cost anything (and might even save money) like letting the grass grow longer and reducing the mowed grass areas. Other things, like planting trees and shrubs or building a storm water retention pond or wetland for the parking lot runoff, need funding or donations of time, equipment and materials. In addition to your environmental education administrative person, help may be found in environmental consulting firms, especially those that consult on how to improve habitat, local or state conservation organizations, state or federal wildlife managers, etc. For example: in Maryland teachers can apply to at least four different places for money to do schoolyard habitat improvement projects. They can get planning help from a wide range of county, state and federal environmental agencies as well as conservation and citizens action programs and several private consulting firms. There are "how to books" for homes and schools.

2. Take advantage of a rainstorm: put on rain gear and go out to actually observe runoff. If you are lucky enough to get rain on a hot, sunny day (not as likely during the traditional school year), take a thermometer with you and compare the temperature of the rain and the temperature of water running off an asphalt parking lot that has been in the sun. Have the students predict what might happen to the fish and invertebrates in a stream if that water ran directly into it. Summer thunderstorms in urban and suburban areas can devastate streams by dumping very hot water into them.

Activity 11
It's raining, it's pouring

Group Name _____

Words that are vocabulary words that may be new to you are written like THIS. Do this worksheet as a group and discuss the words you may not understand. Their meaning is defined, they are used in context, or they are words from other activities in the curriculum, particularly Activity 2. Remind yourself of what aquatic habitats look like by referring to the habitat cards.

A. Watersheds

Read this as a group and do the following

Water that falls on land may sink into the ground. Or it may run off (flow over the ground) downhill into flowing STREAMS or CREEKS which join to form larger RIVERS. Where rivers are dammed naturally or by humans, PONDS (small) or LAKES (large) form standing bodies of water. Where rivers flow into the OCEAN, BAYS form. Rivers, streams, ponds, and lakes (other than rare salt lakes) have FRESH WATER which is not salty. Oceans have SALT WATER. Where fresh and salt water mingle in a bay or ESTUARY, the mixed water is called BRACKISH.

1. Make a table that displays the kind of water that each HABITAT (kind of place like a stream) has based on what you learned in this paragraph. Think carefully about the order in which the kinds of water are arranged.

2. Have the materials person in your group get the following items from the supplies table:

> aluminum baking pan

> sheet of heavy duty aluminum foil

> 4 plastic cups

> plastic milk jug sprinkler with water

Make a model of the land that water runs off of, that is a model of a WATERSHED. Put the plastic cups upside down in one end of the baking pan. These are MOUNTAINS. Arrange the aluminum foil over the bottom of the pan and the cups so that it looks like the land with mountains, VALLEYS, and a low end opposite the mountains. Use the waterer to make it rain in the mountains. Tip it straight up.

3. Describe what happens to the rain. Where does it end up?

4. Find streams, rivers, ponds, and lakes in your model. Where did they form?

5. Arrange these words in a logical sequence starting with rain:

    ocean, stream, bay, lake, river, rain.

6. Explain why you arranged them the way you did.

B. Groundwater

Read this as a group and do the following:

A watershed is all the LAND that drains or has water run off into one body of water. If the land is not hard, the water may sink into the ground rather than running off of the land. Water that sinks in (INFILTRATES) becomes GROUNDWATER which returns to the surface as SPRINGS which flow into streams or lakes. Springs release water year around, providing a constant source of water for wildlife and humans. Early settlers built their homes next to springs. Today groundwater is pumped to the surface for human use from WELLS. One half of the United States population depends on groundwater for drinking water. Soil or other surfaces that allow water to sink in are called PERMEABLE surfaces. Soil that soaks up water releases it slowly as springs. The soil acts like a sponge. Permeable soil and rocks provide a constant supply of water and reduce flooding.

1. Have a person from your group get the following items:

    disposable turkey roasting tray

    plastic milk sprinkler with water

    clear plastic 1/2 or 1 qt deli container (flat design)

    2 cups aquarium pea gravel (rounded pebbles)

    pump from hand lotion or soft soap bottle

    plastic cup

2. Put the plastic dish in the aluminum pan to protect from spills. Pour the gravel into the deli container while one team member holds the pump upright in the dish so that the gravel holds it. Make it rain on the gravel (which represents soil). Describe what happens to the rain.

3. The top of the water under ground is called the WATER TABLE. Draw the plastic dish and gravel here so that it looks like you sliced through the middle and are looking into it from the side (a cross section). Mark the location of the water table.

4. Describe what happens when you push down on the top of the pump repeatedly and catch the water in a cup. What might this represent? How does the water table change?

C. Read this as a group and discuss it.

The water that flows across the land as RUNOFF promptly reaches streams, lakes or bays. Hard surfaces that produce runoff are referred to as IMPERMEABLE. The more impermeable surfaces in an area, the greater the chance of flooding during hard rain or a fast spring thaw. Asphalt, concrete and roofs are totally impermeable. Soils have a range of permeability based on soil type and the amount of plants growing there. Clay is least permeable, and sandy soils are the most permeable. The greater the vegetation, the more rain sinks in rather than runs off. Bare soil and grass catch less water than shrubs or trees. The way humans change landscapes when they build towns and cities or farms has the unintended outcome of reducing groundwater and increasing runoff.

1. Think about the areas around your home and school. Do you think they are more likely to produce runoff or groundwater? Defend your answer using information from above and your knowledge of your local area.

Activity 11
It's raining, it's pouring

Group Name ___possible answers___

Words that are vocabulary words that may be new to you are written like THIS. Do this worksheet as a group and discuss the words you may not understand. Their meaning is defined, they are used in context, or they are words from other activities in curriculum, particularly Activity 2. Remind yourself of what aquatic habitats look like by referring to the habitat cards.

A. Watersheds

Read this as a group and do the following

Water that falls on land may sink into the ground. Or it may run off (flow over the ground) downhill into flowing STREAMS or CREEKS which join to form larger RIVERS. Where rivers are dammed naturally or by humans, PONDS (small) or LAKES (large) form standing bodies of water. Where rivers flow into the OCEAN, BAYS form. Rivers, streams, ponds, and lakes (other than rare salt lakes) have FRESH WATER which is not salty. Oceans have SALT WATER. Where fresh and salt water mingle in a bay or ESTUARY, the mixed water is called BRACKISH.

1. Make a table that displays the kind of water that each HABITAT (kind of place like a stream) has based on what you learned in this paragraph. Think carefully about the order in which the kinds of water are arranged.

fresh water	brackish water	salt water
stream river creek lake pond	estuary bay	ocean

2. Have the materials person in your group get the following items from the supplies table:

> aluminum baking pan
>
> sheet of heavy duty aluminum foil
>
> 4 plastic cups
>
> plastic milk jug sprinkler with water

Make a model of the land that water runs off of, that is a model of a WATERSHED. Put the plastic cups upside down in one end of the baking pan. These are MOUNTAINS. Arrange the aluminum foil over the bottom of the pan and the cups so that it looks like the land with mountains, VALLEYS, and a low end opposite the mountains. Use the waterer to make it rain in the mountains. Tip it straight up.

86 Living in Water   National Aquarium in Baltimore

3. Describe what happens to the rain. Where does it end up?

It collected in the crinkles (valleys) and ran down to the lowest place. A couple of low spots in the mountains made lakes that caught some of the water.

4. Find streams, rivers, ponds, and lakes in your model. Where did they form?

In the bottoms of valleys.

5. Arrange these words in a logical sequence starting with rain:

ocean, stream, bay, lake, river, rain.

rain   stream   ~~river~~   lake   river   bay   ocean

6. Explain why you arranged them the way you did.

This is the path water flowed but I had a river on both sides of the lake.

B. Groundwater

Read this as a group and do the following:

A watershed is all the LAND that drains or has water run off into one body of water. If the land is not hard, the water may sink into the ground rather than running off of the land. Water that sinks in (INFILTRATES) becomes GROUNDWATER which returns to the surface as SPRINGS which flow into streams or lakes. Springs release water year around, providing a constant source of water for wildlife and humans. Early settlers built their homes next to springs. Today groundwater is pumped to the surface for human use from WELLS. One half of the United States population depends on groundwater for drinking water. Soil or other surfaces that allow water to sink in are called PERMEABLE surfaces. Soil that soaks up water releases it slowly as springs. The soil acts like a sponge. Permeable soil and rocks provide a constant supply of water and reduce flooding.

1. Have a person from your group get the following items:

     disposable turkey roasting tray

     plastic milk sprinkler with water

     clear plastic 1/2 or 1 qt deli container (flat design)

     2 cups aquarium pea gravel (rounded pebbles)

     pump from hand lotion or soft soap bottle

     plastic cup

2. Put the plastic dish in the aluminum pan to protect from spills. Pour the gravel into the deli container while one team member holds the pump upright in the dish so that the gravel holds it. Make it rain on the gravel (which represents soil). Describe what happens to the rain.

The water runs down between the rocks and collects in the bottom of the dish.

3. The top of the water under ground is called the WATER TABLE. Draw the plastic dish and gravel here so that it looks like you sliced through the middle and are looking into it from the side (a cross section). Mark the location of the water table.

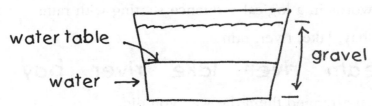

4. Describe what happens when you push down on the top of the pump repeatedly and catch the water in a cup. What might this represent? How does the water table change?

The water comes back up from the bottom and out the pump. It is like an "old time" pump to get water like in the movies. The water table goes down.

C. Read this as a group and discuss it.

The water that flows across the land as RUNOFF promptly reaches streams, lakes or bays. Hard surfaces that produce runoff are referred to as IMPERMEABLE. The more impermeable surfaces in an area, the greater the chance of flooding during hard rain or a fast spring thaw. Asphalt, concrete and roofs are totally impermeable. Soils have a range of permeability based on soil type and the amount of plants growing there. Clay is least permeable, and sandy soils are the most permeable. The greater the vegetation, the more rain sinks in rather than runs off. Bare soil and grass catch less water than shrubs or trees. The way humans change landscapes when they build towns and cities or farms has the unintended outcome of reducing groundwater and increasing runoff.

1. Think about the areas around your home and school. Do you think they are more likely to produce runoff or groundwater? Defend your answer using information from above and your knowledge of your local area.

There is a lot of concrete around our school. Water has to run off just like in the aluminum foil watershed. Our house has grass and trees which catch the rain and hold it until it can sink in like the water did into the gravel and become groundwater.

ACTIVITY 12 How the water gets dirty

What happens to soil and pollutants when it rains? Students use watershed and groundwater models to examine things that dissolve or become suspended in water.

Teacher's information

People typically think of water pollution as something that comes from big factories. Children's books generally blame some nameless company for water quality problems. In fact, federal regulations have made great strides in dealing with many forms of industrial water pollution. The kind of water pollution that has not improved is the kind that each and every one of us contributes to. The solution to many current problems cannot be regulation—no one can watch and monitor all of us! Only public education and changes in our behavior will help.

Water dissolves chemicals that it contacts as it runs over the surface of land and as it sinks into the soil. In the natural world, many of these chemicals are essential mineral nutrients for plants living in the water. Humans have modified the natural world in ways that can greatly increase the amount and kinds of chemicals dissolved in water as well as the soil particles suspended in water. These materials can have a serious impact on things living in water as well as on all the organisms that depend on that water for drinking, including humans. This activity allows students to visualize and model how pollution enters the water.

Water pollution is divided into two categories, depending on how it reaches water: POINT SOURCE and NON-POINT SOURCE. The same chemical can show up as both. For example, nitrate can enter water as a point source from sewage discharge or as a non-point source from lawn fertilizer runoff. Either way, it causes the same problems in the water. Point source pollution has been reduced since the original Clean Water Act because the federal, state and local governments have written and enforced regulations about what can come out of the pipes that are the point sources. A fertilizer factory cannot dump fertilizer into local water as part of the manufacturing process. A farmer who uses too much of the same fertilizer on his/her fields and has it wash away

Science skills
observing
inferring
communicating

Concepts
Water that falls on impermeable surfaces picks up soil particles and chemicals as it runs off, carrying them into streams and rivers.

Water that falls on permeable surfaces dissolves chemicals from the soil and carries them into groundwater.

Many human activities on a watershed lead to water pollution.

Water pollution is divided into non-point and point source categories based on how it enters the water.

Skills practiced
using a model

Time
2 class periods

Mode of instruction
teacher directed group work

Sample objectives
Students make and test models of water pollution.

Students predict which human activities in their own neighborhoods might cause water pollution.

Students propose modifications to their neighborhood that decrease water pollution.

Builds on
Activity 1
Activity 11

Materials

for each group

aluminum turkey roasting pan or plastic cat litter tray

4–5 small plastic cups

30 inch sheet heavy duty aluminum foil

1/2 gal milk jug sprinkler (see Recipes)

clear plastic 1/2 qt or 1 qt deli container (flat design)

2 cups aquarium pea gravel (rounded pebbles)

pump from hand lotion or soft soap bottle

1 pack dark colored powdered drink mix (without sugar)

food coloring in dropper bottle

for each student

worksheet

Preparation

Make the sprinklers if this is the first time you are doing this activity (see Recipes). Duplicate the worksheet.

Outline

before class

1. make milk jug sprinkler

2. copy worksheets

during class

1. discuss student observations of water pollution during a rain storm

2. make watershed and groundwater models

3. use food coloring to make a point source pollution site on each and then rain

4. clean up models

5. use Kool-Aid to make a small non-point source pollution site on each and then rain

6. compare results

during rain into a local stream as non-point source pollution is not regulated. Neither is the home owner who over-fertilizes his lawn. Education is needed to address non-point source pollution.

Introduction

Ask your students to describe ways they have observed water getting dirty when it rains. What kinds of water pollution have they observed? Their answers may range from oil or trash from a city street entering a storm drain to soil washing off a newly plowed farm field in a heavy storm. Try to steer them away from the big factory syndrome and toward local, numerous smaller sources of pollution. Introduce them to the chemical, nitrate, which was discussed in the introduction to this section. They are going to model what happens when too much of this chemical is used on land or is spilled or dumped on land.

Action

Have them get their watershed models or remake them. Do the same with the groundwater model. (See Activity 11 for details.) Challenge them to find out what happens to a chemical that is spilled or dumped from a specific site on their watershed and on their groundwater model. When a pollutant comes from a specific identifiable source that can be measured and regulated, it is called POINT SOURCE pollution. The chemical in question is nitrate. Point sources of nitrate could be a sewage treatment plant outfall, a cattle feed lot or big manure pile, a mismanaged

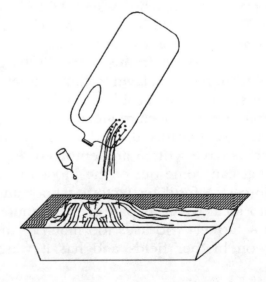

fertilizer plant or storage facility. Have them use a drop or two of food coloring to indicate their point source site for nitrate on their watershed and their groundwater models. Then make it rain. What happens? If you stop the pollution and the water in your bay, lake or river flows away to the sea, will the water get cleaner over time? Carefully, scoop the water out of the model with a cup and make it rain some more. Repeat raining and scooping until the water is somewhat clean. Compare how hard it is to get the nitrate out of the groundwater model by pumping and raining. Groundwater pollution is very persistent!

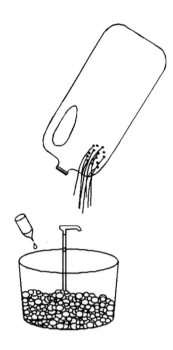

Next, challenge students to describe what they think will happen to nitrate as it is spread across that land as NON-POINT SOURCE pollution. It might be fertilizer spread on a lawn, field, pasture or golf course. It might be manure spread on a field or garden. In cities dog feces can be a major contributor to non-point nitrate pollution. Have the students sprinkle a bit of dark powdered drink mix on the watershed and the groundwater model to represent nitrate. Then make it rain. Where does the nitrate go? If you no longer pollute the groundwater model, how long does it take to get the pollution out? Pump until dry and make it rain again. Repeat until the water is clean.

Results and reflection

Have the students describe their observations. Can any of them relate their own observations of rain flow around their homes to what they observed in their models? Where did the nitrate go in each case? It dissolved in water.

Based on their observations, which would return to normal faster after pollution is stopped—a lake which gets most of its water from runoff or a groundwater-fed spring or well? Lake Erie was once a very polluted lake, but it has gotten much better in the last twenty years or so. Groundwater pollution is much harder to correct.

Ask the students to reflect on the accuracy of their models:

- How are the models of pollution they used in this activity like and unlike real pollution?
- What did you learn about pollution from observing your models?
- What might make these models more realistic or useful?

Conclusions

In either case—runoff or groundwater—it might be easier to prevent pollution than clean it up. No matter what the source, anything that dissolves in water gets into lakes, streams, rivers and groundwater—our sources of drinking water. These chemicals can affect the natural ecosystems and their inhabitants as well as the humans that drink the water.

Extension and applications

1. Have the students make a list of ways that they might prevent non-point source pollution of any kind in their neighborhood. The list could range from not letting car oil drain out on the street or dirt when the oil is changed to fertilizing their lawn less to picking up trash before it washes down the storm drain.

2. Have students write to state or county departments of the environment for information on ways to prevent water pollution around their home, school or neighborhood. Water supply companies or local drinking water departments may also have information. Check your phone book for addresses under city, county and state government or look them up on the Internet.

Activity 12
How the water gets dirty

Name _____

A. Point source pollution (food coloring)

1. Before making a point source pollution site on your watershed model and your groundwater model, describe what you want it to represent and how it might have gotten there in each case.

   a. watershed model

   b. groundwater model

2. When you made it rain, describe what happened to the pollution and where it went.

   a. watershed model

   b. groundwater model

B. Non-point source pollution (powdered drink mix)

1. Before making a non-point source pollution site on your watershed model and your groundwater model, tell what the drink mix represents and describe a story for how it might have gotten there in each.

   a. watershed model

   b. groundwater model

2. When it rained, describe what happened to the pollution and where it went.

   a. watershed model

   b. groundwater

Activity 12
How the water gets dirty

Name **possible answers**

A. Point source pollution (food coloring)

1. Before making a point source pollution site on your watershed model and your groundwater model, describe what you want it to represent and how it might have gotten there in each case.

   a. watershed model   My food coloring is the pipe on a sewage treatment plant.

   b. groundwater model   My food coloring is a big spill from a fertilizer truck.

2. When you made it rain, describe what happened to the pollution and where it went.

   a. watershed model   It rained and mixed with the sewage and ran down to the ocean.

   b. groundwater model   It mixed with the rain and went down and polluted the water in the gravel.

B. Non-point source pollution (powdered drink mix)

1. Before making a non-point source pollution site on your watershed model and your groundwater model, tell what the drink mix represents and describe a story for how it might have gotten there in each.

   a. watershed model   A farmer spread fertilizer all over his field and so did the golf course.

   b. groundwater model   A homeowner put fertilizer all over his big lawn.

2. When it rained, describe what happened to the pollution and where it went.

   a. watershed model   The fertilizer washed into the rivers and the ocean.

   b. groundwater   The fertilizer got into the groundwater. It was hard to clean it out.

ACTIVITY 13 Home sweet watershed

What is your watershed? Students use maps to locate their nearest stream (or storm drain), identify their watershed address and predict local problems.

Teacher's information

Many curricula include an activity like this one. Most are written with the expectation that you will be able to provide topographical ("topo" in geological slang) maps—maps with common elevations shown as lines called contour lines. Topo maps are great, but you may not have them. City, county, state and national street maps usually show streams, rivers, lakes, bays and the ocean. It is better to do this with less perfect maps than to not do it because you do not have topo maps.

Introduction

Review what happened to water in the model watershed when it fell on hard surfaces: it ran off into the low places which form streams. The streams joined into larger streams and then into rivers in a pattern that looks much like the branches of a tree, starting at the tip and working back to the trunk. Indeed, one of the old time names for small streams is "branch." Texans speak of drinking branch water.

Action

Challenge your students to follow the path a drop of water would follow if it fell on the school playground. Where would it go? Let the students use the maps that show your location in its largest scale first and propose answers in their groups. If you have laminated the maps, allow students to literally trace the path with water soluble markers. Then lead a group discussion.

If you live in the city, the entry point to the water system may be a storm drain or gutter. In the rainy areas of the United States, schools may have a stream on the property. In the arid areas, it may be that a dry "wash" or "arroyo" or drainage ditch is the start. Try to figure out where the gutter or storm drain leads. What is the name of the body of water it drains

Science skills
observing
measuring

Concepts
Everyone lives in a nested set of watersheds, starting with the land that drains into the local stream.

Skills practiced
using maps

Time
1–2 class periods

Mode of instruction
teacher directed group work

Builds on
Activity 11
Activity 12

Materials

street maps ranging from local to state and national
local and regional topographic maps if available
rulers
optional
laminating supplies
water soluble markers

Preparation

Collect the maps. Tourist bureaus are a great source of maps. Ask friends who hike or are geologically inclined if they have topographic maps you can borrow. Or check at a local mountaineering/camping store. You can also purchase them from the U.S. Geological Survey, but it takes weeks to get them. If you laminate the maps, the students can use water soluble pens to write on them.

Outline

before class
1. collect maps; laminate if possible

during class
1. review watershed model observations

2. locate school and nearest stream on maps

3. trace the path of a water drop from the school to the ocean

4. write watershed address

5. discuss path of pollutant, refer to watershed model

6. discuss the nature of models and value of two different models

7. compare maps and watershed models

into? If your city has storm drains and sewers combined (a situation that is not ideal, but is common), then you must figure out where your sewage treatment plant is and find the name of the stream, river, lake or bay it empties into.

Have the groups use the maps in larger and larger scale to follow that drop of water as it flows down hill to the sea. Write the name of each body of water as they go. For example, a school in Baltimore City might have a drop of water flow to the storm drain which empties into the Gwynn's Falls (a stream), which would then empty into the Patapsco River, which flows into the Chesapeake Bay which leads to the Atlantic Ocean.

Now have the groups write a list that tells the school's watershed address by listing all the watersheds the school is found in. Remind the students that each watershed is all the land that drains into a named body of water and that the land is named for the body of water. The most sensible way to write it is the same way we write home addresses—the most specific information appears first and the most general last. In other words, it is the most local (smallest) first and the largest last. Just follow the path of the water drop! The school above is in:

• the Gywnn's Falls watershed which is a part of
• the Patapsco River watershed which is a part of
• the Chesapeake Bay watershed

Results and reflection

Have the groups compare their addresses. Were they the same? Identify problems with understanding. Discuss what would have happened if students spilled a chemical that dissolved in water on their school yard. Where would it end up?

Ask the students to consider in what way are the maps a model? Locate one map that shows your watershed. Have them describe how five parts of the real watershed are represented on your map. How is the map like and unlike the real watershed it represents? How is the map model of your watershed different than the models you constructed in Activities 11 and 12?

Have the students write a paragraph explaining why it might be helpful for understanding to use different models to represent the same thing. Then have them share their ideas.

Conclusions

Each of us has a watershed home. For example, in the mid-Atlantic region, some of us who live hundreds of miles apart still have a common watershed address—there are 64,000 square miles in the Chesapeake Bay watershed. Each of us has an impact on it, and each is responsible for helping take care of it. We can do this even though we may never go near the Bay by taking care of our own little piece of its watershed close to home.

Extension and applications

1. Have the students locate and contact local and state environmental agencies (check the phone book under city, county and state government) asking for literature on what the ordinary person can do to protect streams, rivers, lakes, or bays. Also contact not-for-profit environmental groups such as Save our Streams, the Nature Conservancy and the Audubon Society as well as regional offices (check under United States government) of federal agencies such as the U.S. Fish and Wildlife Service and the Environmental Protection Agency. An even better way to contact some of these folks is on the Internet. Search by their name. This will provide a wealth of authentic literature and ideas for class projects to help the students' watershed home.

2. For a special project, have students combine the three-dimensional watersheds of Activities 11 and 12 with the realistic scale and detail of your local map. Make a newspaper and paste model of your region on a large cardboard base. Seal it with paint and then paint the forests green, streams blue, fields brown, roads black, etc.

ACTIVITY 14 Oxygen for life
What is one source of the oxygen that is dissolved in water?
Student experiments using dissolved oxygen test kits.

Science skills
observing
measuring
organizing
inferring
experimenting
communicating

Concepts
Some of the oxygen that is in water comes from the air.

The oxygen dissolved in water may vary depending on the conditions of the sample.

Skills practiced
measuring dissolved oxygen
averaging

Time
3 class periods

Mode of instruction
teacher directed group work

Sample objectives
Students follow instructions to test for dissolved oxygen.

Students explain why experimental results from different sources vary.

Builds on
Activity 1
Activity 2

Teacher's information

Measuring dissolved oxygen is difficult, but rewarding. Just successfully following the instructions can be a major thrill for your students. Oxygen is as important to the animals and plants living in water as it is to those living on land. Oxygen in the water that they "breathe" is DISSOLVED OXYGEN gas, not the oxygen atom in the water molecule. Plants may add dissolved oxygen when they photosynthesize. Much of the dissolved oxygen, however, comes from the air and enters the water at the surface where is dissolves. Once in the water, oxygen DIFFUSES slowly in the water, 300,000 times more slowly than it does in air! Because of this, dissolved oxygen amounts may vary significantly from one place to another in aquatic habitats that are not constantly mixed. In this exercise students will prove that oxygen from air does enter or dissolve in water.

The amount of dissolved oxygen measured may be expressed in several different ways. One way is in parts per million (ppm) which is based on the weight of the oxygen gas versus the weight of the water. Oxygen in natural environments ranges from no oxygen (0 ppm) which is lethal for many organisms to 6–10 ppm which is sufficient for most animals to more than 15 ppm in some cases.

Introduction

Have the students fill a transparent plastic cup with cold water early in the day and let it sit on their desks. Observe it occasionally. If you have individual periods, you may want to fill these and let them sit out ahead of time.

Ask the students what they observed as their cup of water sat on the desk. Some of the students should have noticed that tiny bubbles formed on the sides of the cup. What do they think the bubbles are? Air. Where did it come from? It was DISSOLVED in the water. What do animals need that is in air? OXYGEN. The air is 21% oxygen.

Action

How can we prove that oxygen dissolves in water from the air? Distribute the sealed jars. Tell the students that you have treated the water in the jar to remove most of the air. They do not have to trust you. They can test the water for DISSOLVED OXYGEN themselves using a test kit. Review safety when working with chemicals. Make sure they understand that the color changes are INDICATORS of things that cannot be seen themselves but which cause the color changes.

In the first class period teach the students the techniques of testing for dissolved oxygen with tap water. Have the entire class work together under strict supervision, having carefully reviewed safety. Have one group come forward and demonstrate how to do the test under your direct guidance. Show them how to do the key parts, but then let them do it. The student groups then do each step as it is demonstrated. Work through the steps: student reads instruction for demonstrating group, leadership group does step with teacher guiding it while class watches, then each group does just that step. Review what was done at each step.

Have recorder from each group write its results on the board. They all did the same test on the same water. Why didn't they get exactly the same numbers? Review variables introduced such as the possibility that each group did the test in a slightly different way. For example, did each read the syringe the same way? This is why meticulous care in following instructions is important so that each group's results may be used by the whole class. In a real lab, sometimes only one scientist does a test so that human variability would be reduced. Scientists also write detailed instructions so that others can repeat their work exactly in order to check their results.

No matter how perfect, there is usually still some variability. How do scientists deal with this variability? They use a branch of mathematics called statistics. Demonstrate how to average the numbers the students got. Then have them practice averaging several other sets of numbers.

The following day have the students work in groups taking turns using the chemicals in the test kits and following the instructions provided. They may need

Materials

for each group
small clear plastic cup
cold water
pint or 1/2 pt sealed canning jar filled to the top with water boiled for 20 minutes (see Recipes)
dissolved oxygen sample bottle A (see Recipes)
dissolved oxygen sample vial B (see Recipes)
dissolved oxygen syringe (titrator, see Recipes)
aluminum pie pan
dissolved oxygen test instructions (see Recipes)
shared by several groups
dissolved oxygen test kit (see Recipes)
habitat cards from Activity 2
for each child
safety goggles
worksheet

Preparation

This and all other dissolved oxygen exercises are written for the LaMotte Chemical Company dissolved oxygen test kit. This brand was used for two reasons: the results are read in numerical units with one decimal place for additional accuracy, and the kits are the kind most commonly available to teachers in scientific supply catalogs. There are several other good test kits available. You may modify the instructions to use them. The test for dissolved oxygen in these kits is the Winkler method, an old and very accurate one. Neither you nor the students need to understand the chemical transformations in this kit. You do need to understand what the results mean in terms of the natural world. The Recipes section gives you simplified directions for use of this kit. Make the modifications suggested.

After reading the exercise and assembling the materials, practice the oxygen test yourself. Duplicate the worksheets. Boil and "can" water for this exercise at home. (See Recipes for details.) Allow the sealed jars to cool. Canning jars with rubber seals are needed. Jars such as mayonnaise jars may break when boiled water is added.

your help to open the sealed jars. A blunt table knife can be used as a lever to pry the lid up. Test immediately after opening. Pour carefully into sample Bottle A. Be careful not to add air by agitating the water. Do the test one step at a time under teacher direction. Have students record results on the board. If students made serious errors that you can identify, you may choose to discuss them and delete their data from the final chart. Make sure they understand that you are not just making sure that things come out right. Explain the nature of their error so that they can be careful to correct it. If you have time and another sealed jar, they might repeat their work. Have the students average their numbers.

The third test can be done the following day or immediately, depending on time. What happens if the sealed water sample is in contact with air? Ask the students how they might test this. Either leave the rest of the boiled sample out open overnight to test the next day or immediately after the first test, pour half the water out of each jar, put the lid on and shake the jar hard. Uncap and recap and shake several times. This speeds both diffusion and mixing. Test the water for dissolved oxygen again.

Results and reflection

Compare the first oxygen measurements from the just opened jars with the second after the water had been in contact with the air. Make a table on the board or an overhead projector. Did everyone get the same results? No. Any time a group of folks do a test, there is variation in the way they do it and in the way they read it. That is one of the reasons that scientific tests must be repeated many times, generally by the same person. Average the results for both the tests of the sealed and exposed samples. What are the differences? The sealed samples were much lower in dissolved oxygen than the samples exposed to air.

If the oxygen needed by animals that live in water comes from air, where would the students predict animals that need lots of oxygen might live? Near the surface. If shaking added oxygen, can the students predict which kinds of aquatic habitats might be high in dissolved oxygen? Have them review the cards. Streams where water tumbles over rocks and ocean beaches where waves crash into the shore are high in dissolved oxygen.

Conclusion

Where did the dissolved oxygen in your sample come from? From the air. Gases from the air dissolve in water. Shaking or splashing speeds up the process.

Using your aquarium

Have your students observe the movement of water in your classroom aquarium. The air lift columns create a current that causes the water to circulate, exposing all of the water to the air and mixing the oxygen. What would happen if there were no circulation? Oxygen would be lower at the bottom if it were being used by some of the tank's inhabitants.

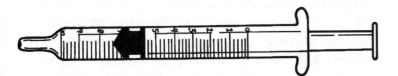

This reads 7.4 ppm.

Outline

before class

1. boil water and seal in small canning jars (see Recipes)

2. copy worksheet

3. copy dissolved oxygen instructions (see Recipes)

during class

1. observe cold water in cups

2. practice the dissolved oxygen test with tap water

3. discuss variation in results

4. test sealed boiled water samples for dissolved oxygen

5. pour out half the sample, cap and shake vigorously or leave out on desk uncovered overnight

6. test sample exposed to air for dissolved oxygen

7. compare differences

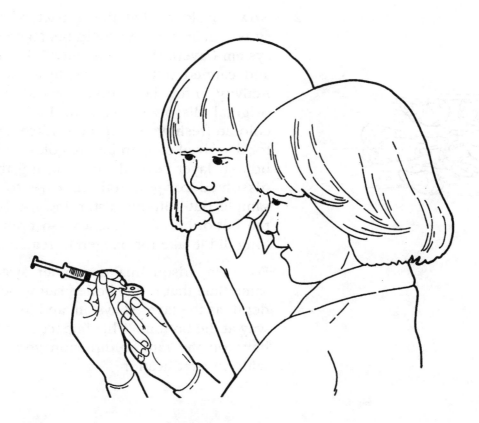

Extension and applications

1. Oxygen normally enters at the surface of the water. Decomposers use oxygen and are frequently located at the bottom of a body of water. Oxygen must enter the water at the surface and diffuse downward. Diffusion in water is slower than diffusion in air. How long does it take for oxygen to get to the bottom of a lake? Try seeing how fast it travels in a fruit jar.

    Uncap a quart fruit jar of boiled water (see Recipes) and let it sit for an hour. First sample the water from right at the surface and then the water from the very bottom. Use a siphon (see Recipes) to take the sample. Do this very carefully so that the water is not stirred.

    You should get a differential, even over the shallow distance in a fruit jar. What does this mean in the real world? If the wind or a current is not stirring up the water in a lake or pond, there will be a good deal more oxygen in the surface water than in the water at the bottom if bottom-dwelling organisms are actively using oxygen.

2. Now consider what the impact of a stratified system in an estuary would be. Make a stratified system (layered) as in Activity 7, but use boiled and canned salt water (made as described in Activity 7) for the bottom layer after testing its original dissolved oxygen. Use oxygenated, colored fresh water for the surface layer. Test it for dissolved oxygen before coloring it. Test the bottom layer several times during the day for dissolved oxygen, taking care to siphon a sample out without disturbing the layers. Test the bottom until it reaches 5–6 ppm DO. How long did it take for oxygen to reach the bottom?

    For comparison, make another system at the same time that is just boiled salt water the same depth as the layered system and test it the same way at the bottom. What impact did the layering have on the rate of diffusion versus diffusion with no stratification?

Activity 14
Oxygen for life

Name _____

1. Describe what you observed as the plastic cup of cold water warmed to room temperature.

2. Record the results of your group's dissolved oxygen test of the sealed jar here.

   a. dissolved oxygen in the water when the jar was first opened: _____ ppm

   b. dissolved oxygen in the water after it was exposed to the air: _____ ppm

3. How much change in dissolved oxygen did you measure in your water sample?

4. Record the results for the class on this table:

	1	2	3	4	5	6	7	8
dissolved oxygen (ppm) in newly opened jar								
dissolved oxygen (ppm) after shaking or sitting open over night								
difference (ppm)								

5. Average difference in ppm _____

6. Explain the difference in dissolved oxygen in the sample before and after it was exposed to the air.

Activity 14
Oxygen for life

Name **possible answers**

1. Describe what you observed as the plastic cup of cold water warmed to room temperature?

   There were tiny bubbles that formed on the inside of the glass. They got bigger as the water warmed up.

2. Record the results of your group's dissolved oxygen test of the sealed jar here.

   a. dissolved oxygen in the water when the jar was first opened: **2.2** ppm

   b. dissolved oxygen in the water after it was exposed to the air: **5.6** ppm

3. How much change in dissolved oxygen did you measure in your water sample?

   5.6
   -2.2   **3.4 ppm change with shaking — there was more!**
   3.4

4. Record the results for the class on this table:

	1	2	3	4	5	6	7	8
dissolved oxygen (ppm) in newly opened jar	2.2	1.8	2.4	2.0	1.6			
dissolved oxygen (ppm) after shaking or sitting open over night	5.6	4.8	5.4	5.2	5.8			
difference (ppm)	3.4	3.0	3.0	3.2	3.2			

5. Average difference in ppm **3.16 ppm**

6. Explain the difference in dissolved oxygen in the sample before and after it was exposed to the air.

   I cannot explain how but oxygen had to get mixed or dissolved in the water from the air when we shook the water. We did not add anything but air.

National Aquarium in Baltimore

ACTIVITY 15 Sour reward!

Acid rain and aquatic environments. What happens when an aquatic environment becomes acidified? Student experiments with aquatic microcosms.

Teacher's information

Some freshwater aquatic environments are very sensitive to the effects of acid rain. Certain rock and soil types contain calcium which buffers the acid, but many do not. For example, half the streams in Maryland are subject to acidification. Acid rain occurs all over the industrialized world. It is the product of combustion of high sulfur fuels (which produces the gas sulfur dioxide) and high temperature combustion of any fuel (which produces the gas nitrous oxide). These gases dissolve in water in the atmosphere, acidifying the rain. These gases can be "scrubbed" from the large factories, incinerators and power plants where they are produced, but not from the small sources like home furnaces, wood stoves and cars. Acid rain also damages terrestrial organisms. Some kinds of plants are particularly susceptible.

Introduction

Start by asking the students about acids—do they have a gut feeling for the word "acid"? Sterile swabs dipped in white vinegar, lemon juice and orange juice can be sampled by a few students (or the whole class) to get an idea of their effects, and the fact that strength of acids vary. **Do not reuse a swab.** Dip, taste and dispose of them. Very lightly touch the tongue with the swab. White vinegar is most unpleasant. Compare touching your skin. Why is the tongue sensitive and the skin not? Humans, birds and reptiles have tough waterproof skin, but the skin on a frog or fish is like the inside of your mouth. Speculate what it would feel like to have vinegar in your eyes. That is what acid water is like for a fish or frog. If you are going to use the pH test kit, let the groups test the pH of tap water, white vinegar, lemon juice, white vinegar diluted with 9 parts water to one part vinegar, and that solution diluted again 9 parts water to 1 part solution of vinegar. It takes a

Science skills

observing
measuring
inferring
experimenting
communicating

Concepts

Some gases such as sulfur dioxide and nitrous oxide that are released into the atmosphere due to human actions also dissolve in water, particularly the water vapor in the air.

Acid rain is produced by these gases in solution.

Acidification of aquatic habitats can have a negative impact on the plants and animals in those environments.

Skills practiced

measuring pH
measuring fluids
using a camera

Time

1 class period set up
1 later class period data collection and analysis

Mode of instruction

teacher directed group work

Sample objectives

Students compare the effects of different levels of acidification on aquatic habitats.

Builds on

Activity 4

Materials

for each large group

4 pint jars or 12 or 16 oz clear plastic
 soda bottles with label removed
 and top cut off
2 c (240 ml) aged tap water with pH
 near 7 (6.8–7.5 range)
1 Nutritab (see Recipes) or uncolored
 plant fertilizer
10 ml plastic graduated cylinder
2 c (500 ml) measuring cup
1/4 c (60) ml white vinegar
2 c (500 ml) rich algal culture (see
 Recipes)
small plastic cup for vinegar
stick-on labels
permanent pen, crayon or pencil

for the class

grow light or window with indirect
 sun

optional

wide range pH test kit and extra test
 vials (see Recipes)
sterile cotton swabs
white vinegar
lemon juice
orange juice
clean disposable paper cups
simple camera with flash
goggles

100 fold dilution before the pH changes much. A third dilution changes pH also.

Ask the students to speculate on the effect of acid on animals and plants living in aquatic environments, based on what it did to their tongues. Where might it come from? In addition to a discussion of acid rain, the students may speculate about industrial sources. Steel mills use an acid bath for their products. Pickle factories use lots of vinegar. Mining can also lead to acid drainage when rain leaches through the disturbed rock.

How could the students test the effect of acidification on an aquatic environment? Help them design this test. **Do not let them use more acid than specified here.** You will be amazed at how fast and effective white vinegar is at killing things. For test organisms, they are going to use freshwater green algae (often referred to as "plants" in that they do photosynthesis, but not classified as plants). In this test they will expose these organisms to several levels of acidification. One sample for each group will not receive any acid. As the control, it will be the thing against which the students can compare their manipulated samples. By doing a range of acid levels, the students can determine if the strength of the acid is important.

Action

Divide into large groups. Collect the materials. Do the preparation step by step all together. The algal samples need to be distributed evenly to each group—shake and pour fast into their cups. Then they must stir before putting into the sample jars. Put 1/2 cup (120 ml) of algal sample in each of the 4 sample jars. Fill each jar to about 1 cup (240 ml) by adding 1/2 cup (120 ml) aged tap water. Label each jar with the group name and test condition. There should be a control and three levels of acidification: Add 1 ml vinegar to the first cup, 5 ml to the second and 10 ml to the third. The fourth gets nothing as it is the control. If you have pH test kits, test and record the acidity of each jar. Put the jars in a window or under a grow light.

Check every day, stirring to suspend the algae and then comparing for color. Always stir from the control to 1 ml, 5 ml and then 10 ml to avoid adding acid to the lower levels. The amount of acid added is not enough to change the results due to different volumes. A camera may be used to record the color differences. What if nothing happened after a week? Was your tap water alkaline? It might have BUFF-ERED the solutions. Add the same amount of acid again.

Results and reflection

Make group posters of the photographs for parents' night. Have each student write a paragraph on the results of the experiment. Trade with another student and critique each other. Trade back and rewrite. Then have a class discussion. Acid kills the algae. Green jars turn clear. The amazing thing is how little it takes! Lead a discussion of the results. Get out the data on your rainfall records from Activity 4. Have you had acid rain in your area? The pH of normal rain is 5.6 due to dissolved carbon dioxide. Where does the acid rain go? The watershed model tells us it either runs off into streams or goes into the ground. Invite students to speculate on what happens to the plants and animals in the streams. Remind them of their experiences in tasting solutions. What will happen to a fish's gills or frog's skin? If you had a test kit, determine the pH of the solutions that turned clear. What pH was deadly to the algae? Have each group select the best paragraph from its students to put on the poster.

Conclusions

Acidification severely impacts many organisms in many fresh water environments.

Using your classroom aquarium

Test the pH of your aquarium before and after a water change. Also, test the pH of the aged tap water you added. Research the ideal pH for your tank in an aquarium handbook.

Preparation

This is a short term project. The results of acidification are obvious in 1 to 3 days. Have the class work in large groups (2 groups minimum, 3 or 4 maximum) to reduce space and container needs. If you are a middle school teacher with a number of sections of the same class, consider doing one set of jars per class and having classes use each other's results.

You need a rich culture of green algae. You can grow one from purchased algal cultures by following the directions that come with cultures from a biological supply house. You can also keep one going by collecting a wild sample from a rich pond and raising it in a big jar (a clean gallon glass or clear plastic jar or a plastic aquarium). You may also be able to start a culture with algae scraped from your classroom aquarium. One of the authors gets her algae by cleaning the bottom of horse and chicken water buckets. Drop in 1/2 tsp of plant fertilizer or a Nutritab every few weeks after removing some water and replacing it with aged tap water. Remove some algae occasionally. Having trouble growing algae? Check the pH of your tap water. One author's pH 6 well water killed both algae and aquarium plants. You may substitute the aquarium plant *Elodea* for the algae in this experiment if you wish, but you need to increase the water in the jars.

Set up the samples of different acidic compounds like lemon juice, if you are doing the optional activity. Use scissors to cut the tops off of 12 or 16 oz clear soda bottles and remove the labels. You may choose to do this experiment with smaller containers and use 1/2 the amounts. If you do so, your students will need to be very careful with their measurements.

Outline

before class

1. purchase or collect algae culture and grow it

2. collect optional acid samples for tasting

during class

1. taste household acid samples

2. test for pH (optional)

3. put 1/2 cup each aged tap water and algal culture in each of 4 jars

4. label and add 1, 5, and 10 ml vinegar to 3; 4th is control

5. photograph and make notes

6. leave in indirect sun or under grow lamp

7. check daily and photograph or make notes

8. discuss results

Extension and applications

1. What are your local streams like? Have students bring water samples from around the school district to test the pH. In some places the water may be very acid. In others, such as the arid Southwest, you may find alkaline water. Here acid rain may hurt forests, but not aquatic habitats because their alkalinity may buffer the acid. In spring, agricultural runoff may be quite alkaline due to lime washing off of pastures and crop land or even golf courses and lawns. If you find acidification (a pH of 6 or lower), contact your local department of the environment or natural resources and ask what is being done. For example, are streams that fish spawn in being treated with lime to correct the pH?

2. What about acid rain into salt water environments? If you have real salt water or "artificial salt water" from the pet store, have the students mix the same amounts as specified above in the algae experiment of acid and water. Was the pH of treated samples of salt water the same as fresh water? Did the pH change the same way? Salt water has a natural buffer system. It should be a pH of 8.3–8.2 and be able to "soak up" some acid before beginning to change. Acid rain is not a problem for estuarine or marine animals and plants. Localized industrial pollution (a steel mill emptying "pickle liquor" from their acid wash) into a brackish system can be severe enough to cause problems.

3. Have students do library or Internet research and report on environmental topics related to acid rain: dying forests, forest soils so depleted of calcium that the trees have stopped growing even though they are not dying, frogs disappearing from lakes. There is also a rich literature in environmental magazines. The students will also find groups which deny that acid rain exists or that it is a problem. Are they able to weigh evidence? Do they have any first hand information? How do they decide which side in an issue is most convincing? Do the students have any first hand experience?

ACTIVITY 16 Dirty water

What happens to nitrate levels in water when soil erodes from the land and enters aquatic habitats? Student experiments with soil and water.

Teacher's information

Moving water moves soil. The stronger the flow of the water, the more soil it moves. Most of the soil settles out of the water as the motion of the water slows or stops. The smallest, lightest soil particles stay up in the water the longest, and the biggest, heaviest settle first. When soil mixes with water, it changes the physical characteristics of the water. Soil particles reduce the penetration of light which may have an effect on aquatic organisms that need light. Soil may clog the gills of the animals that live in water. Soil that settles to the bottom when water movement slows may have an impact on bottom dwelling species. For example, soil may cover fish eggs as it settles. Chemicals from the soil dissolve into the water when they mix. Unlike soil particles, the chemicals cannot be seen, but they can be measured. These chemicals may have an impact on the plants and animals that live in water. In this exercise the chemical measured is nitrate, an essential plant nutrient that can have a major impact on aquatic ecosystems when it is too high.

Introduction

Challenge the students to go out onto the playground and test what happens to the soil when rain falls hard. Can they determine whether plant cover makes a difference when rain hits the soil?

Action

Soil erosion

Give each group the two cards and the "rainmaker" sprinkler. The students are to pick one place where the soil is bare and one where plants such as grass grow. Mark one sail "bare" and one "plants" with a pencil and put each upright in that spot with the card several inches above the ground. Then tip the waterer up straight and make it rain hard from several feet up for several minutes. Return to class. If you have no dirt or

Science skills
observing
measuring
inferring
experimenting
communicating

Concepts
Water moving over bare soil erodes the soil.

Soil mineral nutrients, including nitrate, dissolve in water.

Skills practiced
measuring dissolved nitrate

Time
2–3 class periods

Mode of instruction
independent group work
teacher directed group work
teacher demonstration

Sample objectives
Students compare the impact of vegetation on soil erosion.

Students measure and describe the effects of soil erosion on some of the physical characteristics of water.

Builds on
Activity 1
Activity 12

Materials

for each group

2 3 x 5 inch cards
2 long pencils or thin wooden sticks
1/2 gal milk jug sprinkler (see Recipes)
2 nitrate test vials (see Recipes)
2 each tablets #1 and #2 from nitrate test kit (see Recipes)
10 ml plastic graduated cylinder
2 clean small plastic cups
2 stick-on labels

for class

2 c real dirt or potting soil made with real dirt (not peat moss mix)
8 filter papers or small #2 coffee filters
4 pint fruit jars and lids or 4 large clear plastic jars with lids (such as peanut butter jars)
4 stick-on labels
flashlight
1 or 2 nitrate test kits (see Recipes)
1 qt tap water with one Nutritab (see Recipes)
1 gallon tap water low in nitrate (below 3 ppm)
4 labeled plastic cups

for each student

goggles
worksheet
graph paper

optional

water low in nitrate if the school tap water is over 2–3 ppm nitrogen as nitrate
a pan of dirt and a pan of live grass sod from a lawn if your playground is entirely concrete

grass on the playground, a pan of dirt and a pan of sod dug up from a home lawn will serve as a model.

What did the students observe? What do the sails look like now? What was the effect of the vegetation? The sail on bare dirt should be splashed with mud. The dirt around it should have been picked up by the rain and carried downhill. The sail in vegetation should be relatively clean, and the water sank into the grass beneath it.

Ask if the students have ever seen a creek, pond or river after a very hard rain. Is the water a different color? Can they state why, after the test they just did? The rain running over the surface of the soil has picked up soil particles and carried them into the water. Can the students identify human activities that increase this process, called EROSION? Housing developments, newly logged forests, over-grazing, plowed fields, and tire tracks from off road vehicles can all destroy vegetation and leave soil vulnerable to erosion. What was the effect of vegetation? Plants help hold the soil in place. If you have a stream table, you may also use it to demonstrate erosion.

Light in dirty water

What is the immediate effect of erosion? Have students help you with these preparations. Add dirt or potting soil in the following amounts to 3 labeled pt jars: 2 tbs, 1/4 c and 1 c soil. Leave the 4th jar without soil as the control. Fill with water low in nitrate, cap and shake. The water becomes TURBID as the soil particles become suspended.

What are the short term consequences of soil erosion? Shine a flashlight through each jar onto a white piece of paper. What happens to the light? Would animals and plants be affected by sediment in the water? Plants would have light blocked. Animals could not see. Animals might have their gills clogged. Let the jars sit while you do the next part.

Nitrate in water

How do you measure things you cannot see that get into the water? Practice doing a nitrate test. Have each group get 2 water samples—tap water or tap water with a Nutritab. Each group should also pick up two nitrate sample vials and two each of nitrate test kits tablets #1 and #2 sealed in foil. Have each group test the tap water and the enriched water for dissolved nitrate. Lead them slowly through the nitrate test following the directions in the kit. Later they can follow the printed directions on their own. Record and compare their results.

Nitrate from the soil

The next day check what has happened to the soil in the jars. Is the water more clear than yesterday? Check light transmission and compare with day before. Can the students distinguish which soil particles fell to the bottom and which are on top? The largest should be on the bottom, and the smallest on top.

Carefully pour off some of the water at the top of each jar (including the control with no soil) through 2 nested filters into clean labeled plastic cups that indicate how much soil was in each jar. The #2 coffee filters sit in small plastic cups without needing a funnel. Have groups of students repeat the nitrate test, doing at least two tests of each jar. Record the results.

What if there is more nitrate than the test kit can measure? The vial sample looks like grape juice when you finish the test. Get out a 10 ml cylinder. Put 1 ml of the sample in the cylinder. Add 9 ml of clean tap water (distilled water is better if you have nitrate in your tap water) to make a total of 10 ml. Mix. Pour 5 ml back into a clean sample vial. Repeat the nitrate test. Record the result and multiply by 10. For example, if it now reads 8 then the original was 80 ppm. If it is still off scale, repeat the 1 to 9 dilution on 1 ml of the remaining 5 ml of the first dilution. Now you have diluted it twice so you multiply the result by 10 twice or 100. This gives you the nitrate in ppm. If they have to do a dilution, ask the students to try to explain to you why they have to do it.

Have the students complete a data chart that they design or use the one on the worksheet. Have them combine their results.

Results and reflection

What happened to the soil in the jars? Did the potting soil add nitrate to the water? Was there a relationship

Preparation

Make two small slits in each 3 x 5 inch card or punch holes with the pencil point so that each stick is wearing the card like a sail with the sail at one end. Fill the waterers with water. Mix a Nutritab in a liter of tap water or bottled water. Duplicate the worksheet.

Outline

before class

1. mix Nutritab and water

2. make "sails"

3. copy worksheet

during class

1. test "sails" during "rain" on grass and bare dirt

2. mix soil and water in jars

3. test for light transmission (turbidity)

4. practice nitrate test on tap water and Nutritab water

5. let dirt and water jars sit over night

6. test for turbidity

7. filter water from each jar

8. test each for nitrate

between how much soil was in the jar and how much nitrate was dissolved in the water? If too much nitrate is bad, can your students make suggestions for reducing the possibility that soil and water will mix? What can they apply from their playground observations? Vegetation reduces erosion so anything that reduces runoff over bare soil is good. You may get a discussion of contour plowing from country kids or putting down mulch from city students.

Conclusions

Soil erosion can change the characteristics of water in aquatic environments. It may increase the turbidity of the water, reducing light penetration. Sediment that settles may smother things living on the bottom. A chemical in the soil that dissolves in water may enter aquatic environments.

Using your classroom aquarium

Test the nitrate level in your aquarium. It should be lower than 10 ppm. If it is over 10–15 ppm, you should be doing more frequent water changes. Where does the nitrate come from? Decomposing leftover food, fish urine acted upon by bacteria, and feces all contribute to an increase in nitrate.

Extension and applications

1. Scientists measure the turbidity (cloudiness) of bodies of water with a Secchi (pronounced seki) disk. You can make one for your fruit jar experiment by painting a big metal washer with a pattern of alternating quarters of black and white or use a disc cut from a white plastic disposable plate, marked with waterproof marker and weighted with a metal nut or washer. Attach a cord through the hole and use a permanent marker to mark cm on the cord. You lower the Secchi disk into the water until it just disappears when watching from above. Record the depth in cm. In the field a marine biologist would be recording meters. Murky water might allow you to see 1/2 m while clear water might have 100 m of visibility. Measure the fruit jars after shaking and then periodically afterward to measure the rate at which the sediment settled. Graph the rates.

2. Analyze the labels of the plant fertilizers to discover what some plant nutrients are. Students should find compounds containing nitrogen, phosphate and potassium, three primary plant nutrients. Many brands have a number of other chemicals as well. Test different brands for nitrate.

3. Discuss with the students the common misconception that mineral nutrients are plant "food," that is that plants eat them and get energy just as animals eat food. Mineral nutrients are small bits of matter (molecules or ions) that plants put together using energy from the sun to make larger molecules, such as proteins.

4. Can you identify a problem with soil erosion in your area that affects aquatic environments? If you cannot, you might call your local soil conservation district and ask about a local problem that your class might study. For example, sediment from farms and development contributes to the problems of the Chesapeake Bay. Erosion from logging can smother fish eggs in Pacific Northwest streams. Midwest farms are losing topsoil at an alarming rate. Have the students research problems and propose solutions.

5. Did your students find problem areas of erosion on your playground? Have them design projects to improve the situation. Are there groups that fund such projects in your state? Write grant proposals to get funding to repair your own environment.

Activity 16
Dirty water

Name _____

A. Soil erosion

1. Describe what happened to the paper "sails" when you put them near the ground and made it "rain" with the sprinkler.

   a. paper above bare ground

   b. paper above grass, mulch or other ground cover

B. Light in dirty water

1. When you shine light through the water samples with dirt and the control, what differences did you observe in the amount of light that was transmitted (that passed through)?

2. If you were a plant that needed light to grow, which jar would you prefer to live in? Explain your reasoning.

C. Nitrate in water

1. Record the results of the nitrate test on water samples here:

   a. tap water _____ ppm

   b. nitrate enriched tap water _____ ppm

National Aquarium in Baltimore

D. Nitrate from soil

1. Record the nitrate measurements in each jar made by the class. Each group should test at least two samples:

group number	nitrate (ppm) in the control	nitrate (ppm) in jar with 2 tbs soil	nitrate (ppm) in jar with 1/4 c soil	nitrate (ppm) in jar with 1 c soil
Average				

2. State the relationship between the amount of soil added and the nitrate that dissolved into the water in each jar. How was the amount of dirt added (1, 2 and 8 times as much), related to the nitrate measured?

3. Graph the relationship between amount of soil and nitrate level measured, using the group averages on a separate piece of graph paper. The amount of soil added should be on the bottom axis of the graph. Think carefully about how you will arrange the numbers. Review your graph with your group members. Attach it to these pages.

4. You have just bought a new home with a big yard and a stream at the bottom of the yard. When you test the stream, you find it has 10 ppm nitrate—too high! Describe two things you could do, based on what you have learned in this experiment, to reduce the amount of nitrate from your yard that might enter the stream.

    1.

    2.

Activity 16
Dirty water

A. Soil erosion

1. Describe what happened to the paper "sails" when you put them near the ground and made it "rain" with the sprinkler.

   a. paper above bare ground

   This paper is spotted with muddy brown spots of dirt that splashed up with rain water.

   b. paper above grass, mulch or other ground cover

   This paper is crinkled from getting wet but it is clean. No dirt was moved by the rain which seemed to just disappear in the grass.

B. Light in dirty water

1. When you shine light through the water samples with dirt and the control, what differences did you observe in the amount of light that was transmitted (that passed through)?

   The light passed through the clear jar very well but the muddy water stopped almost all the light.

2. If you were a plant that needed light to grow, which jar would you prefer to live in? Explain your reasoning.

   If I needed light, I'd want to be in the water that the light could pass through so it could reach me — the clean water.

C. Nitrate in water

1. Record the results of the nitrate test on water samples here:

   a. tap water ___2___ ppm

   b. nitrate enriched tap water ___11___ ppm

   depends on area and nitrate pollution

D. Nitrate from soil

1. Record the nitrate measurements in each jar made by the class. Each group should test at least two samples: **these results vary widely with soil used**

group number	nitrate (ppm) in the control	nitrate (ppm) in jar with 2 tbs soil	nitrate (ppm) in jar with 1/4 c soil	nitrate (ppm) in jar with 1 c soil
1	2	4	6	15+
2	2	3	7	15+
3	2	4	6	15+
Average	2	3.67	6.3	15+

2. State the relationship between the amount of soil added and the nitrate that dissolved into the water in each jar. How was the amount of dirt added (1, 2 and 8 times as much), related to the nitrate measured?

There was a direct relationship — the more dirt, the more nitrate.

3. Graph the relationship between amount of soil and nitrate level measured, using the group averages on a separate piece of graph paper. The amount of soil added should be on the bottom axis of the graph. Think carefully about how you will arrange the numbers. Review your graph with your group members. Attach it to these pages.

4. You have just bought a new home with a big yard and a stream at the bottom of the yard. When you test the stream, you find it has 10 ppm nitrate—too high! Describe two things you could do, based on what you have learned in this experiment, to reduce the amount of nitrate from your yard that might enter the stream.

1. **I could make sure dirt did not wash into the stream.**

2. **The Nutritab was fertilizer. I could be very careful with fertilizer.**

ACTIVITY 17 Too much of a good thing

What happens to aquatic habitats when they are enriched with nitrate? A long term student experiment with aquatic microcosms: model aquatic environments.

Science skills

observing
inferring
experimenting
communicating

Concepts

Plant nutrients increase the growth of algae in aquatic habitats.

Small models of aquatic habitats called microcosms can be used to test the effects of nutrients on larger systems.

Skills practiced

measuring nitrate
using a camera

Time

1 class period set up
1 class period results
periodic checks over several weeks

Mode of instruction

teacher directed group work

Sample objectives

Students compare and describe the results of nutrient enrichment on aquatic systems.

Builds on

Activity 16

Teacher's information

Model "ponds" in clear jars reveal what happens when plant nutrients or fertilizer wash off the land and into an aquatic habitat. The NUTRIENTS over-fertilize the pond, causing abnormal growth of ALGAE.

Introduction

Remind the students of what they did during Activity 16. What was the immediate effect of erosion? The water became TURBID as the soil particles were suspended. What substance did we test for that went into solution from the soil? Nitrate. What will the nitrate from the soil do to the algae living in the water it enters? How would we do a test to find out? You can give them the experiment as it is written here or you can work with the students to design one like it. The level of your students may determine how you approach this.

Can the students devise a controlled experiment to test the question? Show them the materials you have. Have each student spend 5 minutes making notes on his/her ideas. Then discuss the ideas for 5 minutes with a partner. Finally, hold a class meeting. Write the student ideas on the blackboard. Diagram the experiment, labeling the parts. Have them identify the controlled variables (amount of algae, light, temperature) and the manipulated variable (nitrate concentration). Can the students formulate a hypothesis for what might happen? What outcomes would support their hypothesis? Negate it?

Action

Here is one plan that works. Each group labels four jars, bottles or cups: tap water or bottled water (control), 10 ppm nitrate, 20 ppm nitrate and 40 ppm nitrate. Add aquarium water with algae or pond water with algae to each jar. Use 1/2 c (100 ml) for a pint bottle, 1/4 c (50 ml) for a 9 oz plastic glass. Add 1 c of the solution that matches the label to each (or 1/2 c for the 9 oz glasses). Mark the water level in each with a marking pen. Why did one jar just get tap water? Ask them. It is the CONTROL against which the other jars are measured. The control gets no treatment and differs by just one VARIABLE from each of the tests. (Students may test the control and the sample jars for nitrate and record the results to make sure of the starting levels in each. The actual measurement of nitrate may be different than the amounts listed here due to variations in Nutritabs, however, the relative differences should be the same. Correct the amounts listed on the worksheet.) The manipulated variable here is nitrate.

Set all four jars where there is good light. Keep the sets from each group together. Do not place them in a location that gets very cold or in hot, direct light. Observe them over the next several weeks, recording changes two to three times each week starting today by photographing the jars side-by-side in good light from close-up. Since algal samples may settle, stir each jar with a spoon before taking the photograph. You may keep the water level at the original height by filling each sample back up to the marked level with distilled water. Write the date on a piece of paper that shows in the photograph and make sure the labels show. Keep the samples in the same place in each picture. At the end, again test nitrate in each jar and compare with original levels.

Results and reflection

What happened? The exact results will vary depending on your algal source and growing conditions. Arrange the photographs in order. Over the weeks, the jars with increasing amounts of fertilizer should show a much more luxurious growth of algae than the nitrate free water. The color may be more blue green than the lower level samples. Why? Plant growth was facilitated by chemicals that stay in the water when soil comes into contact with water. They are collectively called PLANT NUTRIENTS. They are chemicals that are in SOLUTION in the water. The fertilizer we added increased these plant nutrients. Nitrate is the nutrient we tested. Was there a differ-

Materials

for each very large group (two minimum)

4 identical clear containers (plastic soda bottles, canning jars or drink cups)

4 stick-on labels

for class (two very large groups)

1 qt algae culture from a freshwater aquarium, pond or purchased algal culture (see Recipes)

1/2 gal aged tap water (use low nitrate bottled water where tap water has measurable nitrate)

1/2 gal aged tap water with 2 Nutritabs in solution, labeled 10 ppm nitrate

1/2 gal aged tap water with 4 Nutritabs in solution, labeled 20 ppm nitrate

1/2 gal aged tap water with 8 Nutritabs in solution, labeled 40 ppm nitrate

5 milk jugs or large soda bottles with labels for above solutions

indirect sun light or grow light

for each student

worksheet

optional

camera and roll of print film (35 mm or Polaroid best)

nitrate test kit

goggles

1 qt distilled water

Preparation

This activity takes several weeks to complete. It is set up during one class, checked periodically and then completed and broken down during a class. If you have several periods of students, have each class do only one set of jars. The amounts listed above are enough for 4 sets of pint or 1/2 liter containers or 8 sets of 9 oz plastic glasses. If your tap water or well is over 2–3 ppm nitrate, use tested bottled water for the first sample, aged tap water for the second with no tablet added, 1 tablet in the third sample and 3 in the fourth. If your tap water is 8 ppm, the third sample will be 18, the fourth 38 ppm. Be sure to test the bottled water (at the store check the label for nitrate, too). There are no government controls on the chemicals in bottled water. In the past, one author has had good luck with uncarbonated Poland Springs water. Carbonated water is too acidic.

Data collection in this experiment is difficult due to subjective evaluation. Using a camera to record changes solves this problem. You may also challenge students to find another way of measuring algal growth. One group tried reading different size type through the sample. The smaller the type you could see, the less algae. To be scientific each test should be done twice to insure that the results can be repeated.

ence in the amount of growth with the different fertilizer levels? Compare all with the control. Though we mistakenly call these nutrients "plant food" plants do not eat them, but rather use them as building blocks for molecules like proteins. Did the nitrate levels in the jars decline? Have students speculate on where the chemicals went. They are in the algae cells now as part of much larger compounds. Have groups make posters of the photographs and write an explanation of the experiment (question, methods, results and conclusions) for parents' night.

You may be surprised at the amount of growth in the control. There may have been plant nutrients in the algal culture. Also, many water supplies are polluted with fertilizer or plant nutrients from farms, golf courses, manure, human sewage and other sources. That was why you tested your tap water and used bottled water if it had nitrate.

Conclusions

Soil erosion and overuse of fertilizers can cause serious problems in aquatic environments. A little algae growth is good. Too much algae can cause serious problems, however. In over-fertilized conditions, the kinds of algae that grow may change to undesirable, toxic or foul-smelling species. Also, when the algae die, the bacteria and fungi that feed on them may use up much of the oxygen in the system during decomposition.

Using your classroom aquarium

Compare your classroom aquarium with these small environments. How are they the same? Different? Where do the nutrients that support algal growth on the walls of your aquarium come from? They come from waste products such as fish urine and feces and the decomposition of uneaten food. Think about what is used as fertilizer in organic gardens. Animal manure. Where did the animals get the nutrients in the wastes? From the food they ate which directly or indirectly came from plants. Nutrients CYCLE between plants and animals.

Extension and applications

1. Can your students devise any other ways of measuring how much algae growth there was? They could strain each of the jars through a coffee filter or large filter paper and compare how green the different filters got. Label each with pencil first. Could they devise a test that used the mini Secchi disks from Activity 16 Extension?

2. Study the label of the plant fertilizer container to discover what some plant nutrients are. Students should find compounds containing nitrogen, phosphate and potassium, three primary plant nutrients. Many brands have a number of other chemicals as well.

3. Test the effect of fertilizer on green plants. Use the Nutritabs mixed up at the same concentrations as here. Compare plants grown from seeds in clean sand and watered with tap water or each of the three nutrient solutions. Which grew best? Does that mean you should always use more fertilizer?

Outline
before class

1. make Nutritab solutions (see Recipes)

2. copy worksheet

during class

1. label jars and add algae culture

2. add nitrate solution to match label

3. optional test for nitrate

4. place in indirect sun or under grow light

5. check and record or photograph daily

6. optional refill to original level with distilled water

Activity 17
Too much of a good thing

Name _____

1. Arrange all the pictures or descriptions in order from the first date to the last. Study the changes you can observe over time in each. Describe them here:

   control

   10 ppm nitrate

   20 ppm nitrate

   40 ppm nitrate

2. Write a statement that states the relationship between nitrate level and algae growth in your containers.

3. Discuss. Why did one jar get no nitrate?

4. If you were in charge of a golf course that had ponds that grew too much algae and your golfers were complaining, what might you do to fix the problem?

Activity 17
Too much of a good thing

Name _possible answers_

1. Arrange all the pictures or descriptions in order from the first date to the last. Study the changes you can observe over time in each. Describe them here:

control

This jar stayed about the same. They did not die, but the algae did not get greener.

10 ppm nitrate

These algae got much more abundant and the jar turned real nice green.

20 ppm nitrate

This jar got even greener.

40 ppm nitrate

This jar was about the same as the 20 ppm jar — green looks almost blue green.

2. Write a statement that states the relationship between nitrate level and algae growth in your containers.

As nitrate increased, so did algae growth. There was a direct correlation.

3. Discuss. Why did one jar got no nitrate?

We needed to have a control against which to measure the changes in the other jars.

4. If you were in charge of a golf course that had ponds that grew too much algae and your golfers were complaining, what might you do to fix the problem?

Since the Nutritabs are fertilizer if I wanted less algae, I would reduce the fertilizer and maybe I could add something that eats algae.

ACTIVITY 18   What's in our water?

Do the amounts of nitrate in natural bodies of water and drinking water vary? Students test and compare water samples.

Science skills
observing
measuring
organizing

Concepts
Nitrate levels vary in time and space in natural bodies of water and in drinking water supplies.

Nitrate levels correlate with land use practices.

Skills practiced
reading maps
measuring nitrate

Time
1 period

Mode of instruction
independent group work

Sample objectives
Students collect samples and test for nitrate levels.

Students compare different sites for nitrate levels.

Students speculate about land use and nitrate levels.

Builds on
Activity 11
Activity 12
Activity 16
Activity 17

Teacher's information

Nitrate levels in water supplies vary in time and space. Nitrate is of concern for two reasons. First, in aquatic habitats, enrichment with plant nutrients like nitrate changes the abundance and species composition of algae. If the algae species composition changes, then other organisms may change. Species diversity may decline. Some animals may no longer have the algae they need for food. Also, the algae that flourish in over-fertilized waters may grow so thickly that they shade aquatic plants. An over-abundance of algae often leads humans to react by using chemicals which clear the water, but do nothing to solve the underlying problem of over-fertilization.

When the algae use up all the plant nutrients, they die and sink to the bottom. There the dead algae are "eaten" by bacteria and fungi—DECOMPOSERS. Many of these organisms use oxygen in order to do decomposition. The dissolved oxygen levels drop, leaving no oxygen for the animals that live in the water. Fish kills from low oxygen levels are common in over-fertilized (eutrophic) bodies of water. Nitrate levels testable with these test kits are of concern in natural bodies of water. Clean water should give no more than "trace" readings.

In high concentrations, nitrate can affect human health, particularly that of unborn fetuses and young babies. In order for a fetus to get oxygen from its mother, it has to have HEMOGLOBIN in its blood with a higher affinity for oxygen than adult human hemoglobin. Hemoglobin is an iron-containing compound in our blood that carries oxygen. Fetal hemoglobin is replaced by the adult form as genes switch on and off after the baby is born. Fetal hemoglobin combines irreversibly with nitrite, an ion that differs from nitrate by having one fewer oxygen atom, NO_2^-.

When nitrate is consumed in food or beverage, some of the nitrate is changed to nitrite in the stomach and absorbed into the blood. Chemically speaking, stomachs are reducing environments, low in oxygen. Ruminants such as sheep and cows are particularly low in oxygen and produce much higher levels of nitrite than humans with the same nitrate levels. They make good "canaries in the mine shaft" when looking for nitrate pollution in water as they suffer fetal and infant deaths long before humans. No human deaths from nitrate pollution have occurred in the United States, but babies have died in Europe. Nitrate levels in some areas of the United States have affected fetal and baby sheep and cows. Because of health concerns, the Environmental Protection Agency limits nitrate levels in public drinking water supplies to 10 ppm (parts per million) nitrogen as nitrate. (The test kits measure nitrogen as nitrate to match the EPA regulations.) Private wells are not regulated, though some areas have monitoring or educational testing programs.

Action

Is all water the same with regard to nitrate levels? Challenge the students to test and map nitrate levels. They should be able to follow the written instructions if they have done the tests before. Working independently in groups, have them test samples and write the results on the board with the sample name and location. Determine whether each sample was from drinking water or a natural habitat. Then put the results on colored paper and pin it to the site where the sample came from on a map. Use one color (green perhaps) for 0–3 ppm nitrate, another (yellow?) for 4–9 ppm, and a third (red?) for 10 ppm or more. The first indicates no human health concern. The second is reason to think about nitrate. The third is over drinking water standards set by the EPA.

Results and reflections

If there are problems, the map should show possible patterns of pollution. Reflect on student observations of water pollution with watershed and groundwater models. Where might nitrate pollution come from? Is it point source or non-point source? Both. No matter what the source, it ends up dissolved in water. The Environmental Protection Agency regulates pub-

Materials

for each group
one or more water samples
nitrate test vials (see Recipes)
tablets #1 and #2 from test kit (see Recipes)

for class
1–2 nitrate test kits
map of geographic area sampled
three colors of paper

for each student
goggles
worksheet

Preparation

Pick which kind of water samples you are going to test and find a map to match. It could be anything from a county map to a world map if you buy fancy foreign bottled water.

Nitrate is stable in water samples stored out of the light where algae cannot grow. You can collect water samples from streams, lakes and drinking water supplies while traveling on vacation. Your colleagues or students can bring samples from home. It is especially interesting to compare rural well water from an agricultural area with city tap water. If all the teachers and all the students in your school have the same drinking water source (you all live in the middle of a big city), then you can buy bottled water samples from different geographic regions and compare them with your tap water. Many bottled waters are carbonated. If you cannot avoid them, open the bottles or cans several days ahead of time and let the gas out. Shaking helps. The water must be "flat." Do your students correspond with students from another region? Trade samples. You might also make a monthly collection of water samples from your tap water supply which may vary widely with rainfall and season. For example, nitrate might be low in the winter and high in the spring when farmers fertilize and runoff carries it into rivers. Or put a call out for water samples on the Internet. Offer to trade. Be sure that you have exact location, date and whether it is drinking water or a natural body of water on each label.

lic drinking water supplies. They do not regulate private wells nor do they regulate bottled water. Public water supplies are not allowed to have more than 10 ppm nitrogen as nitrate because of possible human health issues. How did your samples measure up?

Did any of the drinking water samples exceed 10 ppm nitrate? Where did they come from? Can the students explain why this might have happened? Well water from agricultural areas is often high in nitrate. Students should be able to explain why if they have done most of the activities in this section, integrating nitrate in fertilizer, dissolving, groundwater, etc. Can they explain very low levels? For example, Poland Spring bottled water is very low. Maine has lots of forests where this water is bottled. Can students apply the lessons of Activity 17 and predict levels of algal growth likely to occur in natural water tested? Low levels, not measurable with these test kits, are best for natural water.

Conclusions

Your students may conclude that your area has good water quality with regard to nitrate. Or they may find it is not good at all. If there are problems, consider doing an extension.

Extension and applications

1. Did any of your samples test above 10 ppm nitrogen as nitrate? If so, make sure the proper authorities know. Look in your phone book under county or city government for the agency responsible for water quality. It may be called the department of the environment. Call information for the city or county if you cannot find the address. Or try the Internet. Many counties have extensive Internet information on their agencies. Have the students write letters reporting their findings and asking if there are programs in place to improve water quality.

2. Research the sources of nitrate in drinking water in your area by contacting local water suppliers, agricultural agents, environmental organizations, etc. Collect information and share it with each other. Have student groups write informational material specific to your area and distribute it to parents and others who can help reduce nitrate use. Have students talk to schoolyard maintenance people. Do they use fertilizer? How do they decide when and how much? Do they test the soil to see if it is needed or do they just use it? Are there programs that people should know about? For example, Maryland offers farmers a nutrient management program that helps reduce farm costs and nutrient pollution. Is there a watershed protection program for your drinking water supply if it comes from surface water? Do you have a groundwater protection program if it comes from wells? These kinds of actions might qualify for service learning in your district.

3. Correspond by email or letters with other groups of students and compare your test measurements.

Outline

before class

1. collect water samples

2. pick maps

3. copy worksheet

during class

1. test samples

2. record and share data

3. locate sites on maps

4. discuss; teacher adds nitrate information

5. make maps with colored flags showing nitrate levels

6. review watersheds and groundwater models and discuss nitrate sources

Activity 18
What's in our water?

Do the amounts of nitrate in natural bodies of water and drinking water vary?

1. Working together, your group should test the water samples you have for nitrogen as nitrate. Follow the instructions carefully. Be sure you keep the samples straight if you have more than one to test.

2. Describe how your group made sure the samples were not switched or confused with each other.

3. Write your results on the board to share with others in your class. Make a table of the class's results showing location, date of sample (if available), nitrate level and whether it was from drinking water or a natural body of water.

source and location	date	nitrate (ppm N as NO3)	drinking water or natural habitat

4. If 10 ppm is the most allowed in drinking water, did you identify any problems with water for humans? If so, where?

5. If anything over a trace (up to 1 ppm) may be a problem in natural bodies of water, have you identified any problems in natural habitats? If so, where?

Activity 18
What's in our water?

Name __possible answers__

Do the amounts of nitrate in natural bodies of water and drinking water vary?

1. Working together, your group should test the water samples you have for nitrogen as nitrate. Follow the instructions carefully. Be sure you keep the samples straight if you have more than one to test.

2. Describe how your group made sure the samples were not switched or confused with each other.

    We each labeled our jars and then we only worked with our own sample. We could not label the test tubes because you have to see through them.

3. Write your results on the board to share with others in your class. Make a table of the class's results showing location, date of sample (if available), nitrate level and whether it was from drinking water or a natural body of water.

source and location	date	nitrate (ppm N as NO3)	drinking water or natural habitat
tap Baltimore City	5/10/97	3 ppm	drinking
creek Baltimore Co.	5/6/97	5 ppm	natural
well Baltimore Co.	5/9/97	15 ppm	drinking
well Baltimore Co.	5/5/97	4 ppm	drinking
Baltimore Harbor	5/10/97	trace	natural

4. If 10 ppm is the most allowed in drinking water, did you identify any problems with water for humans? If so, where?

    Yes, one person had a well that was over EPA limits. She thought since she lived in a nice area in the country she had good water. She was shocked.

5. If anything over a trace (up to 1 ppm) may be a problem in natural bodies of water, have you identified any problems in natural habitats? If so, where?

    The creek in Baltimore County had algae growing on the bottom, and it was too high.

ACTIVITY 19 Water pollution detectives

Interpreting nitrate data: student simulation of a real water monitoring program.

Science skills

measuring
organizing
inferring
communicating

Concepts

Environmental scientists work in groups to discover, describe and resolve water quality problems.

Skills practiced

measuring nitrate
averaging
graphing

Time

2–3 periods
home work

Mode of instruction

independent group work
class collaboration

Sample objectives

Students work in groups collaboratively without teacher direction.
Students carry out a complex project that examines a model of an environmental problem.

Builds on

Activity 11,12, 13, 16, 17, 18

Materials

for each student
Nitrate Pollution information sheet
worksheet
goggles
for each large group
Monroe River watershed map
2–4 nitrate test kit vials (see Recipes)
8 each of tablets #1 and # 2 from
    nitrate test kit (see Recipes)
4 labeled water samples (about 50 ml
    of each) in clean cups (see Activity
    19 Preparation)
4 clean plastic spoons
4 stick-on labels

Teacher's information

Groups of scientists work at widely separated locations to examine and resolve large scale environmental problems. Also, scientists routinely take samples over time and space for later analysis. Nitrate is stable so this activity is a reasonable scenario. If you have one class, different groups within your class can work as if they were groups in different locations. If you have more than one period of science, groups from each class may do the same part of the project, and the different classes may function as if they were in different locations, requiring communication among the periods. Different schools might even collaborate, each doing one part and sharing information on the Internet.

This project assumes your students have done the preceding sections of *Living in Water* on watersheds and on nitrate. It assesses their integration of these activities into a coherent set of information. You need at least 3 groups of students to complete this exercise. Each group gets four water samples representing the four seasons from one location along the mythical Monroe River. The students test their samples and then account for the results of their tests based on the activities they have done on watersheds and nitrate and on the Nitrate Pollution information sheet.

Introduction

Students need technical information on nitrate pollution in order to complete this assessment. The Nitrate Pollution information sheet should be used as a homework reading assignment. Copies should also be available for group reference during the project.

Action

Challenge your students to work as real environmental scientists do on a large scale (geographically speaking) problem. Assign a location to each group:

- Farmtown—the watershed is heavily fertilized corn fields; the corn is fed to cows, chickens and hogs which produce manure.
- Jacksonburg—where the Jackson River enters the Monroe River; its watershed is heavily forested wilderness areas and a national park.
- Lincoln City—where millions of people use the Monroe River for drinking water from an upriver intake and do sewage disposal downriver.

Each group is located in a different county along a large river and works for its county government. All the cities are located on the Monroe River because they were founded before trains and roads when almost everything traveled by water, up and down the river. Each group is responsible for testing the river at their location (marked with an X) on the map for nitrate throughout the year. The samples have been stored, and each group will test a full year's samples at one time. In order to understand the river as a whole, the groups must work together even though they work for different county agencies. They should plan their work before beginning. The Monroe River map gives them details about their watershed and sampling location. Pass out the maps and data sheets. Stand back and let them work it out.

Group	season written on sample cup			
letter on cup	spring	summer	fall	winter
F-Farmtown	sample 4	sample 3	sample 2	sample 2
J-Jacksonburg	sample 2	sample 2	sample 1	sample 1
L-Lincoln City	sample 3	sample 3	sample 2	sample 2

for the class
2 nitrate test kits
4 Nutritabs
4 clean sample bottles: plastic milk jugs or soda bottles
4 stick-on labels
1 gal distilled water from grocery store or tap water lower than 1 ppm nitrate

Preparation
Measure the nitrate in your tap water first. If your tap water is more than 1 ppm nitrogen as nitrate, purchase distilled water. Gallons are available at grocery or drug stores (for use in steam irons). Amounts needed depend on how many classes and groups you have. One quart of each test solution should do 20 samples or more. Each test uses only 5 ml.

Make the water samples in labeled clean bottles according to this recipe:
sample 1—1 liter or quart tap water (below 1 ppm) or distilled water
sample 2—1 liter or quart tap or distilled water with 1/2 Nutritab
sample 3—1 liter or quart tap or distilled water with 1 Nutritab
sample 4—1 liter or quart tap or distilled water with 2 Nutritabs
Prepare student samples by labeling four cups for each group with F, J or L and then one of the seasons. For example, one group gets F spring, F summer, F fall and F winter. Line the cups up to match the table below and put about 50 ml of the sample as indicated in this chart in each.

Outline
before class
1. copy worksheets, maps, and reading sheets

2. prepare solutions and samples

3. assign reading as homework

during class
1. set the scenario

2. distribute worksheets and samples

3. allow students to work, following worksheet directions

4. discuss results

Nitrate Pollution

Nitrate produced by humans

Humans have increased the amount of nitrogen in natural systems. Nitrogen enters in the form of ammonium, nitrate, or nitrogen oxide gas which forms acid rain as nitric acid. Naturally occurring bacteria convert all of these to nitrate. There are also bacteria that fix atmospheric nitrogen in a form available for plant use. These bacteria live in association with the roots of legume crops (alfalfa, beans, clover). They also add to the world's nitrate supply. Our current production of nitrate far exceeds the ability of the natural world to use it. Each year we add more.

Some nitrate sources are at specific sites which can be measured and regulated by laws. They come from a point source. Examples include nitrate in city sewage discharge, nitrous oxide gases from an electrical power station which burns fossil fuels, nitrous oxide from a large trash incinerator, nitrate in manure runoff from a major stockyard, or nitrate or ammonium leakage from a fertilizer plant. Nitrous oxide from cars can be regarded as point source, too.

Other nitrate sources are widespread over entire watersheds (non-point source) and are very hard to regulate by laws. These include nitrate and ammonium from inorganic fertilizer or animal manure spread over fields, lawns, gardens, golf courses and parks, from manure of domestic animals, including both farm animals and pets, and from human waste in septic systems which add nitrate to groundwater. In these instances, education, tax incentives and other programs may be the only solution.

Nitrate in natural ecosystems

Pants and algae need nitrate, but too much nitrate in aquatic habitats causes the too much algae growth. These algae cannot be eaten fast enough by grazers. The algae sink and die. Bacteria which breakdown the dead algae use oxygen. The bacteria use so much oxygen that the bottom water cannot support animals which need oxygen to live.

High nitrate also changes which kinds of algae grow in an aquatic environment. It may favor less nutritious or even toxic species. Places like high nitrate farm ponds, animal watering troughs, and swimming pools all suffer from bad algae due to nitrate.

Sometimes the algae is so dense that it blocks light from reaching plants rooted on the bottom, killing them by shading. Coral reefs become overgrown with algae in nutrient rich water. Corals have single-celled algae living inside the coral animal that need light. Corals die when their tiny helpers are shaded by seaweed growth.

Nitrate in drinking water

Drinking water with nitrate above certain levels is dangerous for both humans and animals. It causes a condition in which the red blood cells of babies and baby animals are unable to carry oxygen. Nitrate is relatively non-toxic. However, it is changed by stomach bacteria to nitrite. In normal adults less than 5% of nitrate taken in becomes nitrite. Adults with low stomach acid and bacterial infections can change as much as 50% of the nitrate to nitrite. Babies make more nitrite because their stomachs are low acid. Babies have special oxygen-carrying chemicals in their blood which combine permanently with nitrite, making them unable to carry oxygen. Babies drinking nitrate-contaminated milk (from their mother or from formula) may die or suffer brain damage from low oxygen. No babies have died yet in the United States, but they have in Europe. Baby cows and sheep are even more likely to be hurt. They have died in some places in the United States due to nitrate water pollution.

Because of human risk, the U.S. Environmental Protection Agency (EPA) limits nitrate in public drinking water supplies to less than 10 milligrams/liter nitrogen as nitrate (10 ppm). No one regulates private wells. They are the greatest risk.

Nitrate itself does not cause cancer, but nitrite can combine with other chemicals to form probable human carcinogens. Farmers exposed to high nitrate in their well water are being studied to see if high drinking water levels of nitrate are linked to cancer.

No easy solutions

What can be done? The EPA regulates point sources, but education and individual action are necessary to reduce many of the non-point sources. Here are some problems and approaches:

Farm nutrient management programs

Farmers may spread commercial fertilizer, manure or both without measuring soil nitrate or timing the fertilizer to coincide with maximum crop growth. This leads to groundwater contamination and surface runoff. Nutrient management programs help farmers plan their fertilizer and manure use. These also save farmers money. Some areas have so much livestock, they lack enough land to dispose of the manure. Composting manure for home or garden use may help.

Urban sewage systems

These systems discharge large amounts of dilute nitrate into surface waters. They are regulated and have to meet standards. New treatment methods reduce nitrate discharge. Some small communities use wetlands they have created for sewage treatment. Bacteria in wetlands are capable of returning nitrogen to the atmosphere as a gas.

Land application of sewage sludge

Sewage treatment removes some nitrate in the solids collected as sludge. This creates a sludge disposal problem. It may be used as fertilizer on farms in place of chemical fertilizer. If it is spread too thickly, sludge nitrate may enter surface waters or contaminate ground water. Over-application occurs because farmers are paid by the ton to take sludge.

Homeowners, parks, sod farms and golf courses

Lawn fertilizer is very high in nitrate. Home owners are likely to use too much. This problem is greatest in the northeastern U.S. where 34% of the total fertilizer use is non-farm use. Lawn products also mix pesticides, herbicides and fungicides with fertilizer, resulting in a chemical cocktail. Soil testing and restriction of use to periods of active plant growth help. The best solution is to change the way we manage our yards, school grounds and parks, reducing grass in favor of trees and shrubs. What grass we have should be cut taller to reduce runoff.

Septic systems

Building houses on rural land around cities causes the construction of many septic systems. These leach nitrate to groundwater, often the same water that the new homes use for drinking water. Nitrate is not filtered out by the soil. A family of four contributes about 73 pounds of nitrate per year to the groundwater. Septic system owners need education about their use and problems. Where septic systems are crowded along shorelines, the water itself becomes overloaded with nutrients leaching directly into it. Changing land use practices and constructing sewage treatment systems are two potential solutions to this problem. The first is very difficult due to opposition from developers, and the second is very expensive.

Monroe River Watershed Map

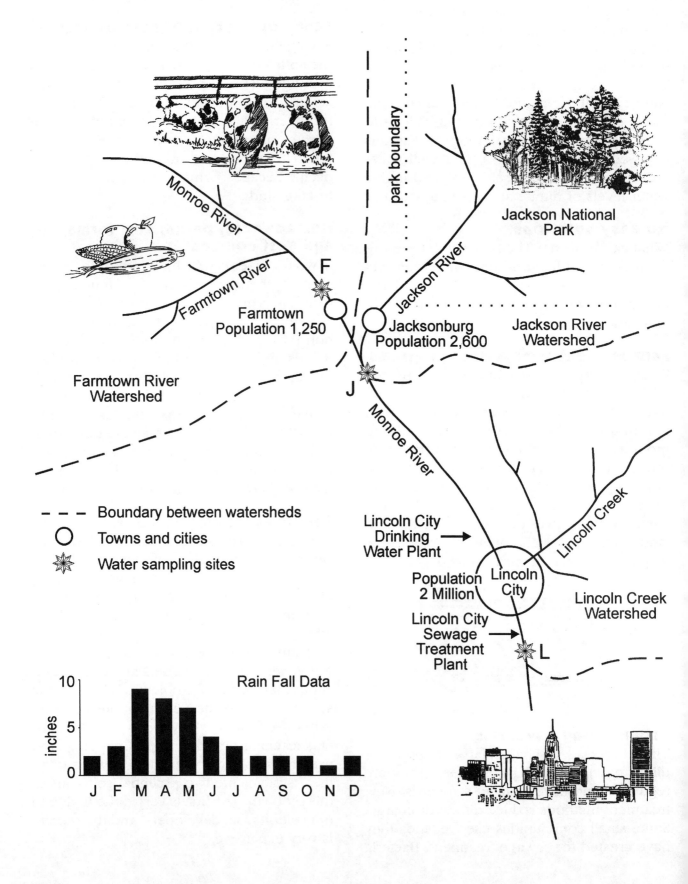

Monroe River

park boundary

Jackson National Park

Farmtown River

Jackson River

F

Farmtown
Population 1,250

Jacksonburg
Population 2,600

Jackson River
Watershed

Farmtown River
Watershed

J

Monroe River

- - - Boundary between watersheds

○ Towns and cities

✳ Water sampling sites

Lincoln City
Drinking
Water Plant

Lincoln Creek

Population
2 Million

Lincoln
City

Lincoln Creek
Watershed

Lincoln City
Sewage
Treatment
Plant

L

Rain Fall Data

inches

10

5

0

J F M A M J J A S O N D

Activity 19
Water pollution detectives

Name _____

A. Working in your group

1. Circle the location of your group: Farmtown   Jacksonburg   Lincoln City

2. What is the major river or stream in your watershed called?

3. Review the Nitrate Pollution information sheet as a group.

4. Using the Monroe River map of your area, the description of your watershed, and the Nitrate Pollution sheet, along with your knowledge of watersheds and water pollution, list the sources of nitrate water pollution you predict your watershed contributes to your sampling location.

a. point source

b. non-point source

5. Test each seasonal water sample twice for nitrate. Design and make a table showing all your data. Also, show the averages for each season.

6. Explain why each test was done twice.

7. Can you account for the nitrate levels you measured by looking only at your own small watershed? Explain your answer.

B. Working with all three groups together

1. Each environmental group sends one person with its data to an annual meeting. Elect your representative and send her/him to the front of the room. The representatives must decide on a table to display all the group information for the whole year in a logical fashion. While they are meeting, your group can clean up. The representatives must present their table to the class. It must be explained and may be modified, based on comments from the class.

2. Draw the final data chart here:

3. Can you now explain the nitrate levels measured in your part of the river better, including why there was seasonal variation? What have your learned by sharing data that helped?

4. Applying information from the Nitrate Pollution information sheet, predict two problems that might occur along the Monroe River below Lincoln City that could be caused by nitrate pollution in the river. They may be problems for things living in the river or for humans.

5. If your group were in charge of the entire Monroe River watershed, list three things that you would do to reduce the level of nitrate in the river below Lincoln City. Explain how each would help improve the water quality of the river.

6. Explain why the rainfall graph was important for understanding the nitrate measurements you got.

Activity 19
Water pollution detectives

Name ___possible answers___

A. Working in your group

1. Circle the location of your group: Farmtown (Jacksonburg) Lincoln City

2. What is the major river or stream in your watershed called?

   ## Jackson River

3. Review the Nitrate Pollution information sheet as a group.

4. Using the Monroe River map of your area, the description of your watershed, and the Nitrate Pollution sheet, along with your knowledge of watersheds and water pollution, list the sources of nitrate water pollution you predict your watershed contributes to your sampling location.

   a. point source   **There might be a sewage treatment plant for Jacksonburg.**

   b. non-point source   **There is almost none except for things like fertilizer in town or nitrogen from air pollution.**

5. Test each seasonal water sample twice for nitrate. Design and make a table showing all your data. Also, show the averages for each season.

season	test 1	test 2
spring	4 ppm	3 ppm
summer	3 ppm	3 ppm
fall	1 ppm	0.5 ppm
winter	1 ppm	1 ppm

6. Explain why each test was done twice.

   We did two to check our work and make sure we didn't make an error.

7. Can you account for the nitrate levels you measured by looking only at your own small watershed? Explain your answer.

   It does not make any sense to have so much as we are a really clean, natural area.

B. Working with all three groups together

1. Each environmental group sends one person with its data to an annual meeting. Elect your representative and send her/him to the front of the room. The representatives must decide on a table to display all the group information for the whole year in a logical fashion. While they are meeting, your group can clean up. The representatives must present their table to the class. It must be explained and may be modified, based on comments from the class.

2. Draw the final data chart here:

	spring	summer	fall	winter
Farmtown	18 ppm	9 ppm	4 ppm	4 ppm
Jacksonburg	4 ppm	4 ppm	1 ppm	1 ppm
Lincoln City	8 ppm	9 ppm	5 ppm	5 ppm

3. Can you now explain the nitrate levels measured in your part of the river better, including why there was seasonal variation? What have your learned by sharing data that helped?

Our really clean watershed added clean water to a polluted river and diluted the pollution. Then the city down river added more nitrate. The big spring rain and fertilizing the fields made a big difference in our results. We could not understand what was happening by looking at one place.

4. Applying information from the Nitrate Pollution information sheet, predict two problems that might occur along the Monroe River below Lincoln City that could be caused by nitrate pollution in the river. They may be problems for things living in the river or for humans.

1. Some day the water might be too high in nitrate for the EPA to let people use it for babies.

2. The algae might grow someplace and get very abundant.

5. If your group were in charge of the entire Monroe River watershed, list three things that you would do to reduce the level of nitrate in the river below Lincoln City. Explain how each would help improve the water quality of the river.

1. I would help the farmers reduce the loss of manure and fertilizer so they could keep it where they need it — on the fields.

2. I would make the Lincoln City sewage treatment plant work really well.

3. I would have programs to teach people how to reduce nitrate pollution in their yards.

6. Explain why the rainfall graph was important for understanding the nitrate measurements you got.

The big spring rain carried soil, manure and fertilizer into the river.

SECTION III Temperature changes and aquatic habitats

Teacher's information

Temperature is very important to living things because the chemical reactions that sustain life occur over a narrow range of temperatures.

Holding heat—specific heat

Have you ever filled a child's wading pool and observed how it heats and cools in comparison to the air temperature? If it is filled with cold water, even in summer it may take several days to warm up enough to play in it. Once it gets warm, it will stay warmer than the air temperature at night. Different substances change temperatures at different RATES even when they are placed in the same conditions. In the example, the substances are air and water. Hence, different materials are said to have different SPECIFIC HEATS. The same amount of heat (energy from the Sun in the example) added to different substances with the same initial temperature will result in different final temperatures.

Water has a very high specific heat. It absorbs a great deal of heat before its temperature rises much. Likewise, water cools very slowly and gives off a great deal of heat when it cools. The gases in air change temperature much more easily: they have lower specific heats. Consequently, aquatic environments do not change temperature very quickly, and they change much more slowly than land habitats. Water is a stable place to live with regard to temperature. The larger the body of water, the more stable it is.

One result of the high specific heat of water is that land masses next to large bodies of water have more moderate climates than those at the same latitude and altitude farther away from water. The water gives off heat during periods when the land is colder and absorbs heat when the land mass is warmer, thus buffering the temperature of the land. You can see this effect by looking up the planting diagram for trees and shrubs given in seed catalogs or in garden books. There are warm bands that extend up both coasts, especially the east coast with its Gulf Stream current that carries warm water up from the tropics to the northern Atlantic Ocean.

Density and temperature

Temperature affects the density of water. Differences in density may result in thermal STRATIFICATION in many bodies of water, with warmer surface waters floating on more dense, colder bottom water. The area of the water where cold and warm water meet is called the THERMOCLINE. It is the region at which the temperature changes rapidly. Density is also affected by salinity with fresh water less dense than salt water. Thus, stratification is also caused by salinity differences. The point at which the saltier water meets the fresh water floating on it is called the HALOCLINE.

Fresh water is most dense, or heaviest, at 4 °C (39 °F). Water colder or warmer than 4 °C floats on the 4 °C water. Because of this, ice forms at the surface of a lake or pond which may remain unfrozen at the bottom during winter, providing a place for aquatic animals to live. The ice may prevent oxygen from entering the water, however.

Density differences also affect the distribution of nutrients and oxygen in water environments. Cold bottom water is frequently nutrient-rich because things that die sink to the bottom and are decomposed there, releasing nutrients like nitrogen and phosphate to the water. The act of decomposition may also deplete bottom water of oxygen. Wind and currents or upwellings may help to mix bottom water with surface water. Bottom water may also reach the surface as surface water cools in fall and its density becomes greater than that of the bottom water. When this happens, surface water sinks and displaces bottom water in a process called TURNOVER. This same process also takes place in the spring as the surface water warms to 4 °C and becomes more dense.

Turnover is often followed by a rapid increase in algal growth as needed nutrients are delivered to the surface where algae are able to do PHOTOSYNTHESIS. In temperate regions, turnover happens at least twice each year. It may occur more than once in a season if the weather warms and cools repeatedly. In tropical regions, lakes, ponds or the ocean are permanently thermally stratified and the surface waters remain nutrient poor.

Temperature and respiration

Water temperature has a direct effect on the plants and animals living in it. Plants and most animals are said to be ECTOTHERMIC (ecto = outside, therm = temperature), meaning that their temperature is determined by the environment. Most aquatic organisms are ectothermic. (The older term "cold-blooded" is inappropriate because these organisms may be warm or cold, depending on their environment.)

Their respiration rates may change with temperature. Thus, their rates of oxygen usage are dependent on their temperature. They use less oxygen when it is cold and more when it is warm because respiration, which is a chemical reaction, goes faster at a warm temperature than a cold one. Increased respiration rates in warm weather may allow greater activity.

Birds, mammals and special members of some other groups are said to be warm-blooded or ENDOTHERMIC. They maintain a constant internal temperature or at least maintain an internal temperature above that of the environment in a cold environment. This constant, warm temperature means that respiration takes place in their cells at the same rate whether it is warm or cold outside. In this case, their oxygen use is greatest in very cold or very hot environments in which they have to do increased respiration to regulate their temperature.

Changing temperatures

Many ectothermic plants and animals adjust their respiration rates over time to be more efficient at the temperature at which they are held. They ACCLIMATE to a specific temperature. The plants you use in experiments may be acclimated at a high or low temperature. This will not negate your results, but may account for high respiration rates at low temperatures if your plants have been held at cold temperatures for a long time. They may be acclimated to the cold.

Animals that live in deep water in the ocean or near the poles experience such constant water temperatures that they have become completely ADAPTED to these temperatures through natural selection and may be killed by a rise of a very few degrees. On the other hand, plants and animals living in smaller bodies of water or in shallow water near shore in larger ones where large temperature changes are common are generally adapted to wide daily or seasonal changes in temperature. Very rapid changes such as those caused by hot runoff from a parking lot may be more than they can tolerate, however. Never heat aquatic animals up as part of an experiment. Organisms living in the hot water in deep vent communities or hot springs are among the few that are adapted to extremely high temperatures that would kill normal organisms. They are of great interest in biotechnology research for this reason.

Gases in solution

The amount of gas in solution in water is directly related to the temperature of the water. Cold water holds far more gas than warm water. Oxygen saturated cold surface water may have well over 10 ppm dissolved oxygen while a warm, shallow tropical lake might have only 6 ppm dissolved oxygen.

The cellular process called RESPIRATION which produces energy in living things generally requires oxygen. However, there are some kinds of animals that are adapted to low oxygen environments such as mud or the low oxygen layer in the ocean called the oxygen minimum layer. Worms that live in mud or turtles that bury themselves in the mud over the winter have special ways of dealing with low oxygen. The worms have blood pigments (like hemoglobin in humans) that are very efficient at picking up what oxygen is available. The turtles go into a special state in which they use very little oxygen. They may even make cellular energy without using oxygen, a process referred to as ANAEROBIC RESPIRATION. A common problem for students is the incorrect use of the term respiration to refer to breathing or gas exchange with the environment.

Most water animals, however, depend on a good supply of oxygen from their environment. When it is not available, they have several options. They may increase the rate at which water passes over their gills. This is like breathing faster when you run because you are using oxygen faster. It is called increased VENTILATION. Another tactic is to move to a better location. In really low oxygen situations, animals like crabs and eels even crawl out of the water. Some animals build up a temporary "oxygen debt" and "repay" it later when more oxygen is available or they are using it less rapidly. Think about running a sprint and then gasping for breath afterwards.

Under natural circumstances, animals generally are not exposed to oxygen stress to which they are not adapted. Occasionally, unusual environmental conditions like prolonged high temperatures may cause abnormally low oxygen in some aquatic habitats. A small pond on a very hot day or night might have low oxygen levels due to heated water and high biological activity (respiration). Frequently, low oxygen levels are caused by human activities. Heating water (thermal pollution) lowers the oxygen it holds. Humans also cause low oxygen in aquatic environments by adding material to the water which will be decayed or decomposed by bacteria. Sewage, plant and animal waste products from food processing plants, animal wastes from farms or feed lots, and organic waste from factories can all serve as food for bacteria living in the water. These bacteria use a great deal of oxygen in respiration as they decompose the wastes. When the problem is compounded by hot water due to climate or thermal pollution, animals that live in the water can die from low oxygen or ANOXIA.

Seasonal changes

Natural temperature changes are a result of seasonal weather changes. There are many organisms which have life cycles keyed to these seasonal changes. MIGRATION of many species for reproduction or feeding is caused by seasonal temperature changes. For example, humpback whales migrate from warm tropical waters where they calve in the winter to colder temperate waters where food is abundant during the spring, summer and fall.

Some fish species migrate seasonally for the purpose of SPAWNING (laying eggs which get fertilized). Most live as adults in the oceans, but enter estuaries and spawn in the estuary or in its rivers. These fish are called ANADROMOUS (from the Greek for running upward). Familiar anadromous fish include salmon, herring, shad, and striped bass. There are also some fish species which do the reverse. They live in fresh water and migrate to salt water to reproduce. American and European eels are the best known members of this group, called CATADROMOUS fish. Migrations are timed to take advantage of specific seasonal changes in water flow, salinity, water temperature and food availability.

As with all species, the total possible number of offspring is much greater than the actual number in each generation because of predation, competition, annual variation in environmental characteristics such as temperature or rainfall and other natural causes of mortality. In the case of some species, the difference between possible and actual numbers of offspring is huge.

In the simulation game about seasonal migration in this section, herring are used as an example of an anadromous fish. Herring occur throughout the northern hemisphere. The herring family includes members which use a complete variety of reproductive strategies. Some species live and spawn at sea. Others may spawn at sea and mature and feed in estuaries (menhaden of eastern U.S. coast) or spawn in freshwater tributaries and migrate to the ocean as adults. It is the last of these, the anadromous herring of the eastern United States, which are used in the game. Blueback herring are also called glut herring for the huge numbers which once glutted the streams each spring. Prior to the arrival of Europeans in North America, these fish existed in incredible numbers. They have been greatly reduced by human actions.

Another anadromous fish, the striped bass, is one of the most prized sportfish on the East Coast. Adult bass in the Chesapeake Bay region spend their winters in deep water in the mid or lower Bay. Since this is a stratified estuary, the deeper water is saltier. It also has more constant temperatures. Larger adults migrate out into the Atlantic Ocean as far north as Nova Scotia. Come late spring (April–June) the adult fish move up the Chesapeake Bay into its tributaries to tidal freshwater areas or only slightly brackish waters to spawn. Strong spring river flow is important to keep the eggs afloat. During the summer season, some striped bass remain in the tributaries, but many move great distances to feed. After the eggs hatch, the larvae migrate downstream as they feed and grow. By winter they join the older fish in deeper water. It appears that these fish, like other anadromous fish such as the salmon, return to spawn each year in the tributary in which they were born. If all the fish from one river are killed, that tributary will not have striped bass again. Salmon throughout the world have similar problems. A combination of overfishing, release of hatchery raised fish, water pollution, sedimentation from development, agriculture and forestry, and dams which block migration can all contribute to the loss of natural populations of wild anadromous fish.

ACTIVITY 20 A change in the weather?

Which changes temperature faster: water or air? Does volume make a difference in how fast a body of water changes temperature? Student experiments with models.

Teacher's information

Much of our weather is dependent on the fact that water absorbs more heat than air for each degree of temperature change. Bodies of water therefore change temperature more slowly and have more stable temperatures than land. Lakes and oceans warm adjacent land in winter and cool it in summer.

Introduction

Hand out the habitat cards for ponds, lakes and oceans. Have the students remind you of their relative sizes. Might size be important in an aquatic habitat? Which is most likely to freeze in winter? What about living in water versus living in air? Ask the students which they think would be warmer on a hot day, a fish living in a big lake or a turtle sitting on a log in the sun next to the lake? How about in the dead of winter when snow is piled up, would it be colder to be under the ice in the pond or sitting on its shore? Generally, the temperature is more constant in water than on land. Have they ever thought about why the climate is more constant under water? Can the students think of test they could do in the classroom that would study how fast temperature changes in air versus water? In a big body of water versus a smaller one? Can we use models to test these questions?

Action

Here is one set of experiments that address these questions. Bring out the cold jars and put in groups on student desks with data sheets. Have students record the time and temperature of each. Leave the jars on the desks. Periodically record time and temperature in each (about every 5 minutes) over the next 45 minutes to an hour. Students may be doing other work during this time, as long as they remember to record their data.

Have the students make a table showing all the class data. Average the results and make a line graph that

Science skills
measuring
organizing
inferring
experimenting
communicating

Concepts
Under the same conditions, water changes temperature more slowly than air.

The larger the volume of water, the more slowly it changes temperature. In terms of temperature, aquatic habitats are more stable than land. habitats.

Skills practiced
measuring temperature
averaging
graphing

Time
2 periods

Mode of instruction
independent group work

Sample objectives
Students compare temperature changes in water and air.

Students graphically display results.

Builds on
Activity 2

Materials

for class

refrigerator or ice chest

cold water

aquatic habitat cards

for each group

3 clear containers (two of same size and one three to four times as big; ideal 1 regular quart and 2 tall 12 oz canning jars; each set of 3 must be the same material

lids or aluminum foil to cover tops of containers

3 thermometers (see Recipes)

for each student

graph paper (see Recipes)

worksheet

Preparation

This activity depends on having access to a cold place: a refrigerator, ice chest or outside on a winter day. If you teach several sections of science, one class may start and others continue this project. This is an opportunity to teach the essential nature of communication in science. Data analysis can be done on the following day.

In selecting containers, do not mix glass and plastic in one set. One group may have an all glass set and another, all plastic. Children should not try to handle large glass containers of water. For safety, you may need to carry the glass containers. Use containers that allow data collection without opening the jars. One good system is slightly off the 1:4 ration of size recommended here. Use tall jelly jars (12 oz) and quart fruit jars (32 oz) because they have good lids and are tall enough for the thermometers. Modify the top of plastic containers to accept the thermometers. The thermometers need to fit inside, but may stick out through holes in foil if necessary. Use cheap, small thermometers that are hard to break, not the long "scientific" glass ones. They should **not** contain mercury (see Recipes).

shows the relationship between temperature and time exposed to the warm room for each size container. Students can also calculate rates of change for each sample by dividing the total temperature change (difference from start to finish) by the total elapsed time (minutes the experiment ran).

Results and reflection

Can the students apply what they learned from their experiment to the real world? What would this mean for you if you lived in water? In air? Generally, animals and plants living in water are subjected to temperature changes that are not as fast or as radical as those that land-living organisms face.

Return to the habitat cards of ponds, lakes and oceans. Lead a discussion about this question. If an animal needs to stay at nearly the same temperature all year, would it prefer to spend the winter and the summer in a big body of water or a little pond? The bigger the body of water, the smaller the changes with season. Can you make any generalizations about the relative seasonal temperature changes likely to be found in a pond, a lake, or the ocean? Small ponds show greater changes in temperature with the seasons. Lakes show less, and oceans even less. But even oceans, at least at the surface, have seasonal temperature changes.

Conclusions

A large body of water changes temperature more slowly than the smaller one. Water temperature changes more slowly than air temperature. Water is a more thermally stable environment than air.

Using your classroom aquarium

Does your classroom experience temperature changes during the night or over the weekend? If you do not have a heater in your aquarium, have your students keep a log of the temperature changes that occur in the aquarium and in the classroom during the day from the time they arrive until they leave. Does the aquarium change temperature much during the day? Compare to the air temperature. If you have a maximum/minimum thermometer, leave it out at night or over the weekend to see how cold the room gets then. Graph the temperature changes of both the room and aquarium on the same chart. Is the amplitude of the change greater in the aquarium or the room? How does this compare with your findings in this activity?

Extension and applications

1. Some students may ask why they did not also study large versus small volumes of air. Let them repeat the experiment, testing the differences in volumes of air.

2. The most famous temperature change on the ocean surface is the periodic El Niño event during which the western tropical Pacific experiences much higher than normal temperatures. This sea surface warming has profound consequences for both sea life and weather all around the Pacific and across North America. Have students research these events and their consequences in the library or on the Internet.

3. Students on the east coast might be interested in studying the impact of the Gulf Stream, a warm ocean current, on their weather. There are exceptional satellite false color images of this current in many publications. They may also be available on the Internet.

Prepare containers the afternoon before class. For each set, fill one of the smaller jars and the larger jar with cold water, leaving a space at the top in case it freezes. Leave the third container filled with air. Add thermometers to each. Put lids on loosely or cover tops with foil. Place in a cold location: in a refrigerator, outside on a cold, but not freezing, night or in an ice chest. Try an ice chest that can come to class unless you have a refrigerator near your room.

Outline
before class
1. copy worksheets and graph paper

2. fill jars and put in cold location

during class
1. use aquatic habitat cards to ask question

2. get jars out and record temperature

3. repeat every 5 minutes for an hour or more

4. compare data

5. graph and compare data

Activity 20
A change in the weather?

Name _____

1. What is the question that this experiment will answer?

2. What evidence would convince you that temperature changed faster in air than in water?

3. The entire class is reading: Fahrenheit _____ Centigrade _____

4. Record the temperatures for your group here

container	Record temperature under time in minutes since start of measurements												
	0	5	10	15	20	25	30	35	40	45	50	55	60
small air													
small water													
large water													

5. Calculate the average temperature for all groups and record here

container	Record temperature under time in minutes since start of measurements												
	0	5	10	15	20	25	30	35	40	45	50	55	60
small air													
small water													
large water													

National Aquarium in Baltimore

6. Use a piece of graph paper to make a line graph of the class results for this experiment. These are the things your graph must have:

- use different colors or different kinds of lines (solid, dots, dashes) to show each kind of sample
- draw a key to which color or style line is which treatment group (i.e. the small air)
- clearly label each axis (each side) with what it is and the kind of units
- put the time on the horizontal axis and the temperature on the vertical axis
- mark the numbers on each axis with the correct spacing
- use a dot to mark the class average at each time interval
- use a vertical bar through each line to show the range of numbers different groups got, showing the highest to the lowest
- connect the dots for each jar with a line

7. Check your graph for the things on the list. Then compare it with the other graphs in your group. Discuss which graph was the best at communicating the class results.

8. List one thing you could change to make your graph better.

9. Now use your graph to answer these questions:

a. Which container changed temperature fastest?

b. Which changed most slowly?

10. Based on your experiment, which of these would you predict would change temperatures through the seasons least? Most? Middle? Write the words in the space.

ocean _____ pond _____ lake _____

Activity 20
A change in the weather?

Name ___possible answers___

1. What is the question that this experiment will answer?

   We are trying to find out if air heats up faster than water. We are also testing to see if how much water is there is important.

2. What evidence would convince you that temperature changed faster in air than in water?

   If the thermometer said the temperature went up faster in the air than the water, I would be convinced.

3. The entire class is reading: Fahrenheit _____ Centigrade _✔_

4. Record the temperatures for your group here

container	Record temperature under time in minutes since start of measurements												
	0	5	10	15	20	25	30	35	40	45	50	55	60
small air	5	9	14.5	18	22	25	25	25	25				
small water	6	8	10	13	16	20	21	23	24				
large water	5	6	9	11	14	17	18	19	21				

5. Calculate the average temperature for all groups and record here

container	Record temperature under time in minutes since start of measurements												
	0	5	10	15	20	25	30	35	40	45	50	55	60
small air	5.2	8.7	14	17.3	22	24.1	25	25					
small water	5.8	7	9.2	12	15.3	19.1	21	22					
large water	5.7	6.2	8	10.6	13	16.1	18.2	19.4					

National Aquarium in Baltimore

6. Use a piece of graph paper to make a line graph of the class results for this experiment. These are the things your graph must have:

- use different colors or different kinds of lines (solid, dots, dashes) to show each kind of sample
- draw a key to which color or style line is which treatment group (i.e. the small air)
- clearly label each axis (each side) with what it is and the kind of units
- put the time on the horizontal axis and the temperature on the vertical axis
- mark the numbers on each axis with the correct spacing
- use a dot to mark the class average at each time interval
- use a vertical bar through each line to show the range of numbers different groups got, showing the highest to the lowest
- connect the dots for each jar with a line

7. Check your graph for the things on the list. Then compare it with the other graphs in your group. Discuss which graph was the best at communicating the class results.

8. List one thing you could change to make your graph better.

I did not make neat lines with the colored pencils and my colors were not different enough.

9. Now use your graph to answer these questions:

a. Which container changed temperature fastest?

air

b. Which changed most slowly?

big water

10. Based on your experiment, which of these would you predict would change temperatures through the seasons least? Most? Middle? Write the words in the space.

ocean __**least**__ pond __**most**__ lake __**middle**__

ACTIVITY 21 Plants use oxygen?

Do the plants and animals that live in water use oxygen? Student experiments measuring dissolved oxygen use by plants.

Science skills

measuring
organizing
inferring
predicting
experimenting
communicating

Concepts

Animals and plants that live in water use oxygen from the water.

Oxygen is used in a process called respiration.

Skills practiced

measuring dissolved oxygen
averaging
weighing
calculating rates

Time

2 periods for plants
2 periods for animals (optional)

Mode of instruction

teacher directed group work

Sample objectives

Students develop experimental evidence that plants (and animals) use oxygen.

Students cite evidence that plants (and animals) use oxygen.

Builds on

Activity 14

Teacher's information

Most people forget that plants constantly use oxygen in RESPIRATION. They think only about the daytime when plants are generally doing enough PHOTO-SYNTHESIS that they produce more oxygen than they use. Most students will swear that plants make oxygen and animals use it. This is a very common misconception.

This activity proves that plants (and animals) use oxygen. They use it in a process called RESPIRATION which takes place in their cells. Plants do respiration all the time, not just at night, but have to be tested in the dark to remove the complication of photosynthesis. Respiration is not breathing in and out (ventilation); it is a cellular process. Like all cellular processes, the rate at which respiration happens is dependent on temperature, so temperature is a variable which must be controlled in this experiment.

This exercise works fine with plants alone. If you are comfortable with experimentation on animals and your school system allows it, they may also be used. Those in your classroom aquarium are ideal. Follow the directions carefully and **do not substitute species.** It is most interesting when several kinds of animals are compared. At least two groups should do each experiment to be able to check for consistent results.

Introduction

Ask your students, "Do plants use oxygen?" Many students will answer no. Do not tell them they are wrong.

National Aquarium in Baltimore

Action

Ask them to design an experiment that will test this question. Show them the jars and plants. They already know how to use the dissolved oxygen test kits. Here is one way to test this question. Test the dissolved oxygen in aged tap water. Slowly, fill a jar to the brim with this aged tap water, place a lid on the jar and label it as the control. To the second jar, add the bunch of *Elodea*. Use about 3 strands 6 inches long per pint. Each jar should be full to the very top, no bubbles. Screw the caps on tightly. (Each large group can divide into two, and each work on one jar at separate tables.) The entire class may use pint jars or 1/2 pint jars with half as much plant material. They should all use the same kind of jars. Ask the students to discuss why they all needed the same kind of jars and the same amount of plants in their jars.

Place both jars in a dark location at room temperature (around 70–75 °F or 25 °C) overnight. The plants must not be in the light. Put them in a cupboard, closet, or grocery bag. If you do this in winter and your school turns the heat way down at night, you might not get very good results as cold temperatures slow the chemical reactions of respiration.

The jars had the same amount of oxygen at the start. Ask the class to predict which jar will have the most oxygen after sitting overnight. The least. Have them give reasons for their predictions. The next day test the water from each of the jars using dissolved oxygen test kits. You may fill the sample bottles by very slowly pouring the water into them over the sink or a pie pan to catch the spills.

Materials

for each large group (2 minimum)
2 pint or 1/2 pint glass canning jars; must **all** be the same size to compare results
3 6 inch strands of healthy *Elodea* per pint (see Recipes)
ruler
aged tap water at room temperature
paper grocery bag or dark place
2 dissolved oxygen titrators or syringes (see Recipes)
2 dissolved oxygen vials B (see Recipes)
2 dissolved oxygen sample bottles A (see Recipes)
2 aluminum pie tins
shared by class
2 dissolved oxygen test kits (see Recipes)
for each student
worksheet
goggles
dissolved oxygen test instructions

Preparation

The materials used in this exercise are the same as Activity 22. Keep the equipment out and do it next. Set out water to age two days ahead of time. Students will set the plant experiment up one day and finish it the next. The plant experiment can be completed in two periods. The optional animal experiment discussed in extensions runs longer than one 45 minute class. If you have the same students all day, it can be completed by one group. Otherwise, start it in one class and finish it in another.

Outline

before class

1. age tap water
2. copy worksheet

during class

1. ask students if plants use oxygen
2. measure dissolved oxygen in tap water
3. put tap water in jars
4. put *Elodea* in one and cap both
5. place in dark over night
6. open and test both for dissolved oxygen
7. calculate change and rate per unit time
8. discuss respiration

Results and reflection

Did the results match the predictions? Remind the students that since all the jars were filled with aged tap water, they started with the same dissolved oxygen levels. Your students should find that the jars with nothing (controls) have the most oxygen and that the levels are lower in the jars with plants. Four jars averaged 0.8 ppm after sitting overnight in an experiment done by children at the Aquarium.

Conclusions

Students who have had lessons on photosynthesis may insist that plants make oxygen, not use it. This experiment should prove that plants (and animals) do respiration and use oxygen from the water in which they live. Do the students think it is important to have good, high levels of dissolved oxygen for animal and plant health? Yes! Without oxygen, the plants and animals will die just as the students would die if deprived of oxygen, which is why we were careful to keep the animals above 4 ppm dissolved oxygen.

National Aquarium in Baltimore

Using your classroom aquarium

The plants and animals in these experiments come from a freshwater aquarium. With careful handling, they will go right back after the experience. Siphon the oil off the surface of the animals' jars before removing them. Following this exercise, discuss the means you use to make sure the oxygen levels remain high in your aquarium.

If you have a large number of plants in your aquarium, test the dissolved oxygen first thing in the morning and compare it with later in the day after the light has come on and the plants have started doing photosynthesis as well as respiration. It should be lower early in the morning since both the plants and animals were using oxygen all night. Have the students speculate on what might happen if the electricity went off during the night.

Extension and applications

1. Optional animal respiration experiment

   Make sure that you handle the animals very carefully. Do not use more of them or different kinds than are recommended. Fish may be transferred in an aquarium net or small plastic bags with water. Do not handle the animals. Practice using the siphon on clean water. Do the siphoning for the students.

   Fill the jars to the very top with the aged tap water after testing it for dissolved oxygen so you know the starting point. Gently place the animal(s) in one jar and use the second as a control. Do not use large animals. One small crayfish (2–3 inch) or one small fish should have a quart jar to itself. A dozen small pond snails can share a jar. Seal the jar and let it sit quietly for about one hour. Do not leave it where students are moving and do not let them tap on it or bother the animal.

   After one hour, use a siphon to remove water and test it for dissolved oxygen. **If the oxygen is below 4 ppm, stop.** A 20 gm crayfish in a one quart jar at room temperature will use enough oxygen to reach 4 ppm in one hour. Snails or small fish will not. If you want to

Materials for extension
for each group
2 quart glass jars with tops
guppy or small goldfish (1 inch) or 1–2 dozen freshwater snails or 1 small crayfish (from pet shop, biological supply house or bait store)
250 gm spring scale with weighing pan or triple beam balance
cooking oil
dissolved oxygen test kits
for teacher
2–3 ft aquarium tubing for siphon (see Recipes)
tubing clamp (see Recipes)
for each student
goggles

continue the experiment, pour about 1/4 inch of cooking oil on the surface of the water to seal it from the air. You cannot just put the lid back on as there will be an air space. Take subsequent samples by putting a siphon below the oil. Siphon or pour the oil off before removing the animal when you are done. Have students devise a data sheet and record their data and

conclusions. Animals also use oxygen. Possible results: control beginning 9.6 ppm, after 1 hour 9.2 ppm; crayfish beginning 9.6 ppm, after 1 hour 4.4 ppm.

2. Quantify your results. Ask your students if it is fair to compare oxygen usage between different species? Can they think of something they could do to make the comparison more fair? Compare oxygen use among different species, including plants, on a weight basis. It is not entirely fair since weight does not take into account the inorganic fraction such as the exoskeleton of the crayfish. You are not about to do what an ecologist would do: cook the animals to inorganic ash to obtain ash free dry weights. Gently shake the plants, snails or crayfish to remove excess water before weighing. Weigh fish in a small bag of water and then weigh the water alone after straining the fish into a net and returning it to its aquarium. Which used more

oxygen per gram of wet weight? Have the students design graphs and charts to detail the differences. Can they identify reasons for differences such as fish using oxygen by swimming? Possible results: Oxygen used per hour by crayfish 5.2 ppm; animal weight 21 grams; animal's use per gram 0.24 ppm/gram.

3. If you take a field trip to a pond with plants, follow the level of dissolved oxygen over a 24 hr period. You should see it go way down at night when everything is using oxygen, and up in the day when plants are adding more oxygen than they use.

Activity 21
Plants use oxygen?

Name _____

1. What question are you going to answer by doing this experiment?

2. What do you predict the answer to this question will be? Give your reasons for this answer.

3. Use this table to record the results of your experiment. Discuss this with your large group members and then fill in the spaces with words that describe which numbers go where.

jars	dissolved oxygen in ppm		

4. Share your data on the board. Then record the class results here as change in dissolved oxygen in each jar. Write the numbers in parts per million (ppm).

	difference for each group (group numbers) ppm				
	1	2	3	4	average
control					
plants					

5. Which changed more, the control or the plants?

6. Calculate the rate at which the plants used dissolved oxygen as amount of oxygen used per unit time.

7. What conclusions about plants and dissolved oxygen can you make based on these results?

Activity 21
Plants use oxygen?

Name __possible answers__

1. What question are you going to answer by doing this experiment?

   I know plants make oxygen but our class is supposed to see if they use it too.

2. What do you predict the answer to this question will be? Give your reasons for this answer.

   I learned in our plants unit in third grade that plants make oxygen. I got good grades. I know that plants do not use oxygen.

3. Use this table to record the results of your experiment. Discuss this with your large group members and then fill in the spaces with words that describe which numbers go where.

jars	dissolved oxygen in ppm		
	oxygen when plants in jar	one day later	change
control—no plants	9.6	9.4	−0.2
plants	9.4	2.6	−6.8

4. Share your data on the board. Then record the class results here as change in dissolved oxygen in each jar. Write the numbers in parts per million (ppm).

	difference for each group (group numbers) ppm				
	1	2	3	4	average
control	−0.2	−0.4	+0.2	—	−0.13
plants	−6.8	−7.2	−6.4	−8	−7.1

5. Which changed more, the control or the plants?

The plants!

6. Calculate the rate at which the plants used dissolved oxygen as amount of oxygen used per unit time.

Time was 24 hours. Average difference was 7.1 ppm

$$\frac{7.1\ ppm}{24hrs} = 0.3\ ppm/hr\ change$$

7. What conclusions about plants and dissolved oxygen can you make based on these results?

The plants really did use oxygen when they were in jars in the dark.

ACTIVITY 22 When the heat's on. . . .

What is the effect of temperature on the rate (amount per unit time) at which aquatic organisms use oxygen? Student experiments measuring dissolved oxygen use.

Teacher's information

This activity examines the relationship between temperature and respiration rates, as measured by dissolved oxygen use, in ECTOTHERMS—living things whose internal temperatures are determined largely by their environment. These include fungi, bacteria, algae, plants and most animals except birds and mammals. Since all other activities in an organism are dependent on the energy which is gained from cellular respiration, this is a very important feature of an organism's life. The term "cold-blooded" has gone out of usage because it conveys misinformation. The organisms are not cold—their temperature tracks that of their environment—nor do many of them have blood. The term ectotherm is self defining. "Ecto" means external and "therm" means heat or temperature. How about animals that move just as fast in winter as in summer, such as dogs, cats and birds? These animals are ENDOTHERMIC which means warmth from inside. They expend energy to regulate their internal body temperature at the same level regardless of the cold or heat.

RESPIRATION is correctly used to refer to a biochemical process that take place in cells. The act of moving air or water across surfaces in order to expose them to maximum oxygen concentrations is VENTILATION, or breathing for things with lungs. The trade of carbon dioxide for oxygen between organism and environment is referred to as GAS EXCHANGE. In this experiment the students are measuring gas exchange as a function of temperature. The amount of oxygen used is a measure of cellular respiration. It is a chemical reaction, and as such, speeds up when the chemicals are warmed up.

Your results will depend to some extent on the temperature at which the plants have been maintained over the preceding weeks. You will have more dramatic differences if the plants have been in a warm (75 °F) aquarium for several weeks. Many

Science skills
measuring
organizing
inferring
experimenting
communicating

Concepts
The temperature of the environment has a direct effect on the rate of oxygen use by living things.

In the case of living things other than birds and mammals, the warmer the environment, the more oxygen is used.

Skills practiced
measuring dissolved oxygen
measuring temperature
averaging

Time
2 periods

Mode of instruction
teacher directed group work

Sample objectives
Students interpret experimental data about the effect of temperature on the rate at which plants use oxygen.

Builds on
Activity 21

Materials

for each of two very large groups

4 pint or 1/2 pint glass canning jars with canning lids, must all be the same size

6 6 inch strands of healthy *Elodea* or 3 for 1/2 pints (see Recipes)

ruler

4 thermometers

aged tap water at room temperature

two brown paper grocery bags

4 dissolved oxygen sample bottles A (see Recipes)

4 dissolved oxygen vials B (see Recipes)

4 dissolved oxygen titrators or syringes (see Recipes)

shared by class

2 dissolved oxygen test kits (see Recipes)

refrigerator or ice chest

for each student

goggles

worksheet

dissolved oxygen test instructions

Preparation

Set the water out ahead of time so that the chlorine will dissipate. Copy worksheets.

ectotherms are able to ACCLIMATE somewhat to prevailing temperatures over time, adjusting their rates of respiration by making different enzymes. Hence a plant can make enough chemically useful energy to survive both during cold periods and very warm times. Some animals, such as trout, also display temperature acclimation of respiration rates. In winter the plants in the cold will use oxygen, though not at the rate of the warmer ones. In summer, those in the cold may use almost none.

Using plants to test the effect of temperature on rates of oxygen usage avoids the possibility of hurting animals. The same kind of work could be done with any aquatic animals, but they may be sensitive to temperature changes. In general, aquatic organisms are much more tolerant of being cooled rapidly than heated quickly.

Introduction

Briefly review the results of Activity 21 in which students proved that things living in water used oxygen from the water. Do the students think that the temperature of the water might affect how fast (the rate at which) plants use oxygen? The concept of rate is a difficult one. It is important to make sure students understand it. You might think of examples of use of the term in more common circumstances such as how fast one eats or how fast a car moves. For example, Juan can eat two hamburgers in 5 minutes while Bob can only eat one in 5 minutes. Which is faster? Can the students express hamburger consumption in units? Rates are expressed as something per unit time. Bob eats 0.2 hamburgers per minute. Ask the students to develop other examples using the speed of cars, how fast they can ride a bike, etc.

Return to the question of temperature and oxygen use. What units would the students use? How about ppm per hour? Can they design a way to answer the question about the effect of temperature? Work with them to modify the experimental design of the previous experiment, Activity 21.

Action

This experiment is similar to Activity 21 in design. Set it up the first day. Test the dissolved oxygen in the aged water and record the amount. Students can set the experiment up in 2 large groups (of 8–12 students each) and divide the testing phase into smaller subgroups (4–6 students), each testing two jars. Each large group should have four jars of aged tap water which has been tested for dissolved oxygen: two controls with no plants and two with aquatic plants. The hard part is to get similar amounts of plant material in each jar. Measure the *Elodea*, putting the same total length in each jar. Three 6 inch pieces in a pint jar is about right. Make sure that the pieces have the same general health as well as length. Fill the jars to overflowing and seal. Put them in the brown paper grocery bags for darkness if needed. Put one control and one jar with plants in each of two locations: cold and room temperature. One of them might be in a refrigerator or in an ice chest or outside on a cold (but not freezing) night and the other at room temperature. Each large group of students will have two jars at room temperature and two in the cold: one test and one control in each. Leave overnight. If your classroom is allowed to get very cold at night, put the warm jars in an ice chest with a gallon jug of very hot water when you leave.

The following class day, have the students test the dissolved oxygen level in each jar. Record the temperature of the water in each after the oxygen sample is taken. Calculate the difference between the test and the control at each temperature. Then compare the amount of oxygen used by plants in the two locations.

Results and reflection

Have your students fill out the worksheets and then discuss their findings. At which temperature was the most oxygen used? There should be a higher rate of use at higher temperatures. Why did we use a control at each temperature? Because temperature could have an effect on dissolved oxygen too, and we wanted to eliminate that as a cause of our results. How might this influence the amount of energy

Outline

before class

1. age tap water

2. copy worksheets and instructions

during class

1. test dissolved oxygen in aged tap water

2. fill 4 jars/group to top with aged tap water

3. label jars: warm control, warm plants, cold control and cold plants

4. put 3 6 inch strands of *Elodea* in plant jars

5. seal all with no bubbles

6. put cold in ice chest or refrigerator

7. put warm at room temperature

8. leave overnight

9. next day unseal and test for dissolved oxygen

10. complete worksheets and discuss

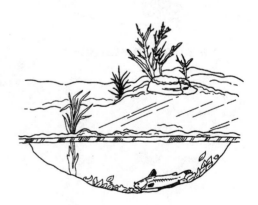

a pond in winter

available to an aquatic plant for growth and reproduction in different seasons?

Can the students think of ways animals that are ectotherms might be able to increase their temperatures? They can move to a warmer spot and often bask in the sun out of water. Turtles and marine iguanas do this.

Conclusions

Environmental temperatures affect the rate at which ectothermic organisms use oxygen and make energy available to do biological work inside themselves. Generally, the warmer the environment, the higher the oxygen usage.

Using your classroom aquarium

Add the *Elodea* to your aquarium. Discuss with your students whether the animals in your aquarium are endotherms or ectotherms. Can they name any endotherms that live in fresh water? Beavers and river otters are two that come to mind. The ocean? Whales, dolphins and seals are groups that live in the ocean.

Extension and applications

1. Discuss the problems animals living in a small pond might have on a very hot day. In a crowded pond, they might exhaust the supply of oxygen. When would the problem be the worst? At the end of the night, because the plants would be using oxygen all night, but not making any with photosynthesis. On the other hand, when would oxygen use be lowest? During the winter when ice might seal the top of the pond from the air, making oxygen less available.

a pond in summer

Activity 22
When the heat's on . . .

Name _____

1. State the question you are trying to answer by doing this experiment.

2. What would you predict the answer might be? What evidence would you need to support this conclusion?

3. The original amount of dissolved oxygen in the water was _____ ppm.

4. The beginning temperature was _____.

5. Record the results of your group's experiment here:

end temperature	dissolved oxygen in ppm			
	beginning	after 24 hrs	difference	amount used per hr
cold plants _____				
cold control _____				
warm plants _____				
warm control _____				

6. Record the results for your class.

jar	dissoved oxygen in ppm					
	1		2		average	
	difference	rate	difference	rate	difference	rate
cold plants						
cold control						
warm plants						
warm control						

7. Based on the class results and discussion in your group, what conclusion can you draw about plants' use of oxygen in the dark with regard to the effect of temperature?

8. Why do you think you added all the class results together and took an average?

9. Can you think of another test you might do to find additional evidence for the effect of temperature on oxygen use by endothermic species? Describe it.

Activity 22
When the heat's on . . .

Name __possible answers__

1. State the question you are trying to answer by doing this experiment.

   We proved plants use oxygen so now we are trying to find out if temperature makes a difference in how much they use.

2. What would you predict the answer might be? What evidence would you need to support this conclusion?

   I think the plants might use more oxygen when it is cold. I eat more in the winter time.

3. The original amount of dissolved oxygen in the water was __9.6__ ppm.

4. The beginning temperature was __24°C__.

5. Record the results of your group's experiment here:

end temperature	dissolved oxygen in ppm			
	beginning	after 24 hrs	difference	amount used per hr
cold plants 5°C	9.6	5.2	–4.4	0.18
cold control 5°C	9.6	9.2	–0.4	0.02
warm plants 24°C	9.6	1.8	–7.8	0.33
warm control 24°C	9.6	9.4	–0.2	0.01

6. Record the results for your class.

jar	dissoved oxygen in ppm					
	1		2		average	
	difference	rate	difference	rate	difference	rate
cold plants	−4.4	0.18	−4.1	0.17	−4.25	0.175
cold control	−0.4	0.02	−0.2	0.01	−0.3	0.015
warm plants	−7.8	0.33	−8.0	0.33	−7.9	0.33
warm control	−0.2	0.01	0.0	—	−0.1	0.005

7. Based on the class results and discussion in your group, what conclusion can you draw about plants' use of oxygen in the dark with regard to the effect of temperature?

The plants in the warm temperature used almost twice as much oxygen in 24 hrs as the plants kept in the refrigerator so in our experiment we can conclude that cold slowed down oxygen use in plants.

8. Why do you think you added all the class results together and took an average?

Averaging smooths out the differences due to minor differences in technique and in things like how much plant we had in the jar.

9. Can you think of another test you might do to find additional evidence for the effect of temperature on oxygen use by endothermic species? Describe it.

We could do the same test on crawfish. We could put some in the refrigerator and leave some at room temperature in jars but we would need to be very careful. It might hurt them.

ACTIVITY 23 When the oxygen goes . . .

What happens to the oxygen dissolved in water when water temperature increases? Student experiments measuring dissolved oxygen.

Teacher's information

The relationship between water temperature and dissolved oxygen is critical for many aquatic animals. The higher the temperature the water is, the lower the amount of gas in solution. This exercise demonstrates this relationship. If you carefully prepare the samples, the students will get surprisingly good results. Samples must be held at the temperatures for a number of hours because gas diffuses slowly through the samples. Only boiling is fast as the sample is stirred by the bubbles.

Introduction

Is there a relationship between the amount of dissolved oxygen in water and the temperature of the water? How can we find out? Can the students help you design a way to test this question? They should suggest heating or cooling water and testing it. Explain how you have treated the water samples ahead of time and sealed those that were heated above room temperature. This is hard for students to understand because you are giving them cooled samples. Ask questions to make sure everyone understands that the amount of oxygen in the heated samples reflects the amount at the temperature it had when sealed. Spend the time to make sure this is clear. How would the students put the samples to use? Test the samples for dissolved oxygen.

Action

The sealed jars may require adult help to open them, using the edge of a table knife. Test the oxygen levels in each water sample and record the results. Do not open until just before using. Be especially careful not to transfer water vigorously as this will add oxygen to the sample. Hold the sample bottle over something to catch the drips and very gently pour the sample down the side of the sample bottle.

Science skills

measuring
organizing
inferring
experimenting
communicating

Concepts

Dissolved oxygen decreases as water temperature increases.

Time

1–2 class periods

Mode of instruction

teacher directed group work

Skills practiced

measuring dissolved oxygen
averaging
graphing

Sample objectives

Students complete an experiment to determine the relationship between water temperature and dissolved oxygen.

Students apply understanding of the relationship of temperature and water to real world problems of aquatic organisms.

Builds on

Activity 14

Materials

for each group

dissolved oxygen sample bottle A
    (see Recipes)
dissolved oxygen vial B (see Recipes)
dissolved oxygen titrator or syringe
    (see Recipes)

shared by class

2 dissolved oxygen test kits (see
    Recipes)
8 pint or 1/2 pint canning jars (see
    Recipes)
two sets of four water samples
    treated as follows (see Recipes)
2 jars water open to air with ice in it
    or in a refrigerator (4 °C)
2 jars water at about 25 °C sitting
    open at room temperature
2 jars water heated to and held at
    about 50 °C for several hours, then
    sealed
2 jars water boiled for 20 minutes
    and sealed
thermometer good to 100 °C such as
    kitchen candy thermometer
table knife to open jars

for each student

goggles
worksheet
graph paper (see Recipes)

If possible, test the water in each jar twice to check the results.

Results and reflection

Let the students fill out the worksheet and then discuss this experiment. The students should get very straightforward results: the higher the temperature to which the water was originally heated, the lower the oxygen in solution. This is a simple relationship based on the physics of gases in solution. When might this be important to an aquatic organism? What might happen to the dissolved oxygen on a hot summer day in a shallow farm pond?

Conclusion

Cold water holds more dissolved oxygen than hot water. What is the consequence of this fact for the animals that live in warm water or in small bodies of water that change temperature? Can the students put this together with the effect of temperature on respiration rates? It means aquatic animals use oxygen the fastest at the same time that it is least available during high temperatures. Fish kills can be caused by nothing more than a long stretch of hot weather.

Using your classroom aquarium

Have your students test the dissolved oxygen in the aquarium water. Record the temperature and the dissolved oxygen. Compare it to the results of this experiment. Does your aquarium have more or less oxygen than the water sample heated to the same temperature? It may have more if it is filled with plants that are doing PHOTOSYNTHESIS. It may have less if it is full of animals or bacteria using oxygen.

Extension and applications

1. Can your students suggest human activities which might result in low dissolved oxygen due to water temperature increases? If electric power generating plants use natural bodies of water for cooling their steam, could the heating of the natural water have an impact on the plants and animals that live in it? Thermal pollution can be

a problem, especially on bodies of water that are relatively small such as rivers and lakes. Dissolved oxygen may get so low that animals die. Because of this, cooling towers are built next to the power plants which allow the heat to be lost to the air before the cooling water is returned to the river or lake. Sometimes special lakes are built which are used only for the power plant cooling.

Preparation

Prepare the water samples. The cold and room temperature samples should be left open at the indicated temperature for 24 hrs. Leave one pair sitting out in the classroom and another pair open in a refrigerator or ice chest. The "canned" heated samples may be prepared at home any time as they are good until opened. Pour the water gently into the jars full to overflowing and put on canning lids. Water sealed in canning jars can cool without picking up additional oxygen. Do not open until ready to use. Do not try to substitute other kinds of jars for canning jars. Label the jars clearly with the treatment temperature.

Outline

before class

1. copy worksheets and graph paper

2. prepare water samples

during class

1. explain to students how samples were prepared

2. open water samples and test for dissolved oxygen

3. complete worksheet and graph

Activity 23
When the oxygen goes . . .

Name _____

1. State the question you want to answer by doing this experiment.

2. Record the results from each group's tests:

temperature (when sealed)	dissolved oxygen in ppm in jars when opened		
	first test	second test	average

3. Use a piece of graph paper to make a line graph showing the average result for each temperature. Use a bar through the line to show range on each point. Compare graphs with other students. Check to make sure that all the parts labeled all the things correctly and marked the units in uniform sections. You may improve your work after discussion with others.

4. Read off of your graph to predict the amount of dissolved oxygen you would expect to find in a sample heated to 75 °C and write it here.

5. What conclusion about the relationship between dissolved oxygen and water temperature can you make based on the results of this experiment?

Activity 23
When the oxygen goes . . .

Name __possible answers__

1. State the question you want to answer by doing this experiment.

   We are trying to find out if there is a relationship between how hot water is and how much dissolved oxygen it can hold.

2. Record the results from each group's tests:

temperature (when sealed)	dissolved oxygen in ppm in jars when opened		
	first test	second test	average
5°C	9.6	10.4	10 ppm
25°C	8.6	8.0	8.3 pmm
50°C	5.6	5.2	5.4 pmm
100°C	1.8	0.8	1.3 pmm

3. Use a piece of graph paper to make a line graph showing the average result for each temperature. Use a bar through the line to show range on each point. Compare graphs with other students. Check to make sure that all the parts labeled all the things correctly and marked the units in uniform sections. You may improve your work after discussion with others.

4. Read off of your graph to predict the amount of dissolved oxygen you would expect to find in a sample heated to 75 °C and write it here.

   3.3 ppm

5. What conclusion about the relationship between dissolved oxygen and water temperature can you make based on the results of this experiment?

   The hotter the water was when it was sealed, the less oxygen it had. The higher the temperature, the lower the dissolved oxygen.

ACTIVITY 24 When the oxygen is gone

How do animals respond to low oxygen environments? A teacher demonstration.

Science skills

observing
inferring
predicting
communicating

Concepts

Two possible animal responses to low oxygen environments are increased movement of water over the gills and migration to higher oxygen habitat.

Animals experiencing low oxygen may be under stress.

Skills practiced

averaging

Time

1 class period

Mode of instruction

teacher demonstration

Sample objectives

Students list ways animals respond to low oxygen environments.

Builds on

Activity 21
Activity 22
Activity 23

Teacher's information

This activity demonstrates several responses of fish to low oxygen levels. It does not take long as the fish behave in obvious ways which are expressed promptly: they do not like doing without oxygen. Use small goldfish from your aquarium or feeder goldfish from the pet store. Goldfish will not be hurt by this experiment. They are very well adapted to low oxygen environments and thrive where other fish die. **Do not substitute other species.** Also goldfish are easier to observe than other species. This is done as a teacher-led demonstration so that you can maintain control of both student and fish behavior.

Introduction

Most aquatic animals get oxygen from the water with GILLS. What would the students do if they were an aquatic animal that gets oxygen from their water environment and, suddenly, there was not enough? How might they test this question? One obvious way is to compare aquatic animals in low and normal oxygen environments and see what differences in behavior can be observed.

How do we make water that is low in oxygen? Review what students learned about water temperature and dissolved oxygen. Water that is heated has little dissolved oxygen, so use boiled water that has been sealed in canning jars. Ask them why the water temperature must be the same (room temperature) in both jars—fish are ectotherms whose oxygen use is directly related to environmental temperature just as it was with plants in Activity 22 so temperature is a controlled variable. Also, hot water might hurt the fish. Students must not tease or otherwise disturb the fish. If they do, the experiment will be invalid as fish will use more oxygen if stressed. They must be quiet and never tap on the jar or otherwise disturb the fish.

Action

Have students stand around a central table, tallest behind and select a team to make observations and record data for everyone. You may use new students for each set of observations on a new fish, but keep the same students for one fish to standardize observations. Select the jar with regular room temperature aged tap water and carefully place a fish in it. After 15 seconds record the number of times it opens and closes its gill covers (opercula) or mouth in a 15 second period. Repeat for a second 15 second period. This is a way to measure how much water it is passing over its gills.

Remove the fish by pouring the water out of the jar through the net and place it in a newly opened jar of room temperature boiled water. Watch it closely. Wait 15 seconds and then count the number of times in 15 seconds that the fish opens its gill covers and record the results. Seeing this and counting accurately are hard and may require practice. Repeat the observation. Do not leave the fish in the water lacking oxygen for more than 3 minutes. Note that the goldfish may exhibit an alternative strategy to increasing its rate of moving water over its gills (ventilation rate). Return the fish to its aquarium home promptly.

Repeat with a second fish and the second set of jars. Have students chart the data on the board. They can convert the gill ventilation counts to number per minute by multiplying by 4.

Alternative action

A non-quantitative way to do this activity is to compare the behaviors of fish placed in water with dissolved oxygen and oxygen depleted water. Put them in at the same time and put the jars side-by-side to compare. Do they show the same responses? Where do they go in the jar? Is one more calm than the other? The fish without oxygen goes to the surface, gulping air. Remove it promptly to the aquarium.

Materials
for class
2 1 inch goldfish (do not substitute other fish)

2 1 quart canning jars filled with water boiled 10 minutes and then sealed; must be at room temperature when used (see Recipes)

2 1 quart canning jars of aged tap water unsealed at room temperature

watches or clock indicating seconds
small fish net
for each student
worksheet

Preparation
Make the low oxygen (boiled water) at home. Boil water in a pan or tea kettle for ten minutes. Then pour gently into two wide mouth quart canning jars. Do not substitute mayonnaise jars which may crack when the hot water is added. Fill to over flowing and put fresh canning lids in place. Tighten the rings using a hot pad to protect your hand. This may be done well ahead of class. The water must be at room temperature when used. Do not substitute fish species. Do not use bigger fish bigger than one inch.

Outline

before class

1. make boiled and aged water

2. copy worksheets

during class

1. ask students to predict fish behavior in low oxygen

2. put each fish in aged water and then in low oxygen

3. count opercular movement in 15 second intervals in each

4. observe behavior in each

5. do worksheets and discuss

Results and reflection

When the students have completed their worksheets or journal entries, lead a discussion of their observations. Fish in low oxygen move water over their gills more rapidly than those in oxygenated water. If the water is very low in oxygen, the only response you may see is the oxygen deprived fish going to the surface and gulping. Goldfish can get oxygen from the air if their gills are wet. Goldfish are well adapted to life in low oxygen habitats.

Conclusions

In these experiments you have looked at several ways animals respond to water which is low in oxygen. One way is to increase the rate at which water moves over the gills. If you have ever watched a crab or shrimp in a small container, you may have noticed increased movement of the water across its gills. This reflects increased ventilation, just as the fish increased movement of water by "swallowing" faster.

A second tactic is to move to an area of the water that has a higher oxygen concentration. In this case the fish moved to the surface and used air. Both responses use energy. Animals that can migrate within an environment are relatively lucky. Can your students name some aquatic animals that cannot move to an area with higher oxygen? Any aquatic animal that is sessile, that grows attached to a surface, cannot move. Oysters, barnacles, mussels, worms that live in tubes are just a few of these. If they cannot move, they may die during periods of unusually low dissolved oxygen.

Using your classroom aquarium

These fish can go into your freshwater aquarium if it is not above 70 degrees Fahrenheit. Goldfish do best in water cooler than that for tropical fish

Extension and applications

1. Can your students identify a local problem caused by low oxygen? Have them research this question by writing or calling one or more local organizations responsible for water quality. Use the phone book or the Internet to find county government environmental departments or a local conservation organization. You may already be aware of such a problem. Once a problem is identified, have your students try to find out when it occurred, the magnitude of the problem, the possible causes and any responses that were made or could be made in the future.

    Common problems are related to cooling power plants or the release of water with a high oxygen demand from sewage plants during warm weather. There are also oxygen depletion problems downstream for paper mills or slaughter houses on occasion. Sometimes the cause is prolonged hot weather, a natural occurrence. The point is to follow the process of the discovery of a problem and the responses to it.

Activity 24
When the oxygen is gone . . .

Name _____

1. Describe the experiment your teacher is going to do.

2. What is the question you hope to answer by doing this?

3. Record the class data here.

   (multiply times moved in 15 seconds by 4 to get number per minute)

	number of times the mouth or gill covers open per minute	
	water with oxygen	water without oxygen
Fish # 1		
Fish # 2		
Average		

4. Why was it important for you to remain quiet and not disturb the fish during this experiment?

5. Describe and compare the behavior of the fish in low dissolved oxygen water with that of the fish in water with normal oxygen levels.

Activity 24
When the oxygen is gone . . .

Name __possible answers__

1. Describe the experiment your teacher is going to do.

   We are going to put goldfish in water that does not have much oxygen and see how they act. They live where the oxygen is low sometimes.

2. What is the question you hope to answer by doing this?

   We want to know what goldfish do to deal with low oxygen.

3. Record the class data here.

   (multiply times moved in 15 seconds by 4 to get number per minute)

	number of times the mouth or gill covers open per minute	
	water with oxygen	water without oxygen
Fish # 1	41	59
Fish # 2	47	71
Average	44	65

4. Why was it important for you to remain quiet and not disturb the fish during this experiment?

   If the fish is stressed by us it would need even more oxygen and it might change the result.

5. Describe and compare the behavior of the fish in low dissolved oxygen water with that of the fish in water with normal oxygen levels.

   The fish in normal water just stayed in the middle of the jar once it got used to the jar. In low oxygen it started pumping water faster and acted nervous. Then it went to the surface and breathed air and relaxed.

ACTIVITY 25 In hot water?
Which weighs more (has greater mass): hot or cold water?
Student experiments with weighing and measuring.

Science skills
observing
measuring
inferring
predicting
experimenting
communicating

Concepts
Water is heaviest (most dense) at temperatures near freezing.

The same volume of water weighs less when hot or when frozen than at 4 °C.

Water of different temperatures may form stratified aquatic systems.

Skills practiced
measuring volume

weighing with a balance, spring scale or triple beam balance

layering water

Time
1–2 periods

Mode of instruction
independent group work

Sample objectives
Students compare the relationship between water temperature and weight per unit volume (density).

Builds on
Activity 6

Teacher's information

This exercise examines the relationship of water temperature to weight (or mass) per unit volume, or density. This relationship has important consequences for aquatic habitats. In the absence of wind, waves or tides, warm water floats above the cooler water without much mixing. Under continued calm conditions, a body of water may become layered or stratified with regard to the temperature of the water. The point at which the warm and cold water meet is called the THERMOCLINE. If you have gone swimming in a lake or pond, you may have experienced these temperature zones.

Animal distribution may be determined by the distribution of different temperatures of water. For instance, trout may concentrate in the lower water of a lake in the summer where it is cooler and, hence, more oxygen-rich. In TEMPERATE regions, seasonal changes in surface water temperatures result in exchanges of surface and bottom water called TURNOVER. Turnover affects the distribution of nutrients and dissolved gases. In the TROPICS where temperatures are consistently warm, permanent thermal stratification has many consequences.

Introduction

Do this activity in two parts: weighing first and then layering with discussion following each part. Tell the students they are going to ask some questions about whether warm or cold water is heavier (has greater mass). Can they actually weigh the two kinds of water and find a difference? What are some ways they could test this question? They should recall Activity 6 and be able to suggest the same techniques. They might suggest weighing the same volumes of warm and cold water. What about warm and cold water? If one is heavier than the other, what would happen when warm water and cold water meet? Can they draw what they predict will happen? Have they done an experiment like this before? How did they address this kind of question with salinity? Some may predict cold water is lighter because ice floats.

Action

Let groups work independently with little direction from you to answer these questions. Weigh the water samples. Things to remember and discuss with groups having problems include: to compare weights, exactly the same volume is needed. If the students do not use the exact same amount of water in each cup, it will not be a fair test. Fill one cup with very hot water (but **not** hot enough to burn a child's skin) and another with refrigerated (cold) water. Measure the water carefully before adding it to each cup. Either measure the cups separately if you have a spring scale or triple beam balance, or put each on one of the two pans of a simple balance. With a large, very accurately measured sample the cold water is heavier if the scales are very accurate. This is difficult for children to do accurately, so you should get different results from groups.

Stop and discuss the results. Students should not get consistent results because the differences in density are very small, smaller than most students' margin of error in weighing and measuring. Discuss the sources of error possible in their work.

How can the students address this question in another way? What happened when they layered salt and fresh water? What would happen if cold and warm water meet in a natural habitat? What would they predict? Would one float on the other? Would they mix? Here is a way to test this. Each group needs two cups of ice water and two of hot water. Add the same color of food coloring to one of the hot and one of the cold water cups. With a plastic spoon very carefully lay a sample of the cold colored water on the surface of the hot clear water cup. Put the bowl of the spoon level with the surface of the water and rotate the bowl out from underneath the sample. What happens? Try the reverse, hot colored water on the surface of the cold clear water cup. What happens? For controls, try the colored water with water of the same temperature. Layering is so precise that

Materials

for each pair of groups to share
1/2 gal hot tap water (120 °F maximum) in a labeled plastic milk jug
1/2 gal cold water (from refrigerator or ice water) in a labeled plastic milk jug

for each group
unbreakable thermometer (see Recipes)
two colors of food coloring
simple balance
250 gm Ohaus spring scale or triple beam balance
2 cup measuring cup
25 ml plastic graduated cylinder
100 ml plastic graduated cylinder
four clear plastic cups (9 oz)
plastic spoon

for each student
worksheet

Preparation

There are a number of different curricula that use different techniques for layering than this does. This one gives very clear results. Practice it before class. If you do not lay the colored water into the uncolored water very gently, it will mix. The key is to very gently rest the spoon with the colored water into the surface of the uncolored water and then rotate the spoon out from under the colored water. There is no pouring. The spoon should just slide out from under the water.

If you lack access to a school refrigerator where you can put jugs of water the night before, use ice cubes in a bucket and scoop out cold water. If you do not have an ice chest for the ice cubes, put them in watertight plastic bags and wrap in newspaper, a wonderful insulator. You need hot tap water from somewhere at school. The temperature should not be high enough to burn students. Many institutions have cool tap water as a safety measure. If so, an alternative is to make hot water with an automatic drip coffee maker which produces hot, but not boiling water. Boiling water is too hot for safety and will melt the plastic half gallon milk jugs.

it will give good results even when you cannot measure the difference with a scale.

Results and reflection

Could the students measure the difference between hot and cold water? Could they tell the difference by layering water? Compare sensitivity of the two tests. Have the students explain why they needed to do the layering test two ways.

Conclusions

Warm water is lighter (less dense) than cold water. When warm and cold water come together and are not stirred, the warm water floats on the cold water. This stratification means that one body of water can have two totally different kinds of places for animals to live in terms of temperature. Lack of mixing can also cause an unequal distribution of things in solution in the water such as oxygen, which diffuses in at the surface or comes from plants at the surface, and nutrients and carbon dioxide, which are released near the bottom as bacteria decompose dead plants and animals.

Stirring mixes the water. What would "stir" water in a real habitat? Wind, which also causes waves, currents, tides or running water would mix warm and cold layers.

Using your classroom aquarium

Looking for a quick way to reinforce the principles of this lab? Fill a small balloon with very cold water (stick a water balloon in the refrigerator) and then drop it into the aquarium. Does it sink or float? Let it stay there during the day. What happens as it warms up? How can this observation be explained in terms of the density of water?

Extension and applications

1. How do temperature stratified systems mix? Let groups of students experiment with their cups of water. Blowing across the surface may be enough to mix them. What happens if an ice cube is added?

2. To demonstrate why ice floats, fill a clear plastic cup with cold water. Mark a line at the top of the water with a marker. Freeze the water and compare with the line. The water expanded as it froze. If the same amount of water takes up more space, then it must be less dense than it was. Ice floats because water expands as it freezes. The water molecules get farther apart. For this reason, water in a crack can break a rock or concrete as it freezes.

3. Water temperature has implications for the mixing or lack of mixing of human sewage from outfalls. If the waste water has a temperature different from the receiving body of water, it will not mix unless there are winds or currents. Many times the engineers assume that waste products will be diluted enough that they are not harmful. Often this does not happen because of temperature differences.

The combined effect of warm, freshwater sewage entering cold salt water is that the sewage floats to the surface. This can be demonstrated. A cup of cold salt water is the ocean. Fill a straw with hot colored fresh water, the waste water. Lower the straw tip to the bottom of the cup and slowly release the water. The warm colored water does not mix, but comes to the surface. Do the waste products always get diluted? No.

Outline
before class
1. copy worksheet

at start of class
1. fill milk jugs with hot and cold water

during class
1. have students attempt to answer the question by weighing samples or comparing with a balance

2. stop and discuss success

3. remind them of previous technique of layering

4. check answers by layering

5. discuss applications to real habitats

Activity 25
In hot water?

1. How did your group try to answer the question, using hot and cold water samples and measuring and weighing devices? Describe what your group did.

2. Compare the results of your first tests weighing hot and cold water with what other groups got. Did you all get the same results? List two reasons why you might have gotten different answers.

3. Describe what your group did when you layered colored water.

4. Draw the results of your layering experiment here and label the parts.

5. Which test do you feel was more accurate? Give evidence for your answer.

6. As a result of your observations with hot and cold water, where would you expect to find the coolest water in a lake in the summer time?

7. If you went diving in a deep water submersible in the ocean, what would you predict would happen to the temperature outside the sub as you descended?

8. You used a model in this activity. Explain how it was like the real thing and how it was different.

9. How was using a model helpful in understanding how a lake or pond might work with regard to warm and cold water?

Activity 25
In hot water?

Name <u>**possible answers**</u>

1. How did your group try to answer the question, using hot and cold water samples and measuring and weighing devices? Describe what your group did.

   We tried to weigh the difference between the same amount of hot and cold water. We also used a balance to see if we could tell the difference. We remembered about big sample sizes and we weighed and measured very carefully.

2. Compare the results of your first tests weighing hot and cold water with what other groups got. Did you all get the same results? List two reasons why you might have gotten different answers.

   We didn't all get the same answer. Maybe there isn't any difference.

   1. We might have not been careful enough.

   2. Our tools might not be accurate enough.

3. Describe what your group did when you layered colored water.

   We used food coloring to mark the water and did both tests to make sure food coloring didn't make a difference.

4. Draw the results of your layering experiment here and label the parts.

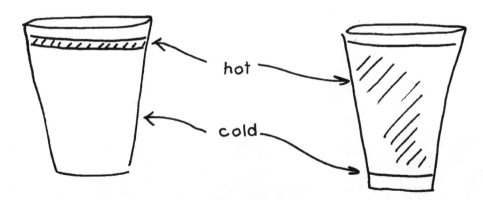

hot

cold

5. Which test do you feel was more accurate? Give evidence for your answer.

The layering because everyone got the same results.

6. As a result of your observations with hot and cold water, where would you expect to find the coolest water in a lake in the summer time?

In the summer it should be colder on the bottom of the lake.

7. If you went diving in a deep water submersible in the ocean, what would you predict would happen to the temperature outside the sub as you descended?

I think the temperature of the water would get lower.

8. You used a model in this activity. Explain how it was like the real thing and how it was different.

The cups let us test this question but they were very small compared to a real lake. They did allow us to experiment. I'm not sure the water would ever be that hot in the real world. We could not put food coloring in a lake to see the layers of water.

9. How was using a model helpful in understanding how a lake or pond might work with regard to warm and cold water?

We could do tests that we could not do with a real lake. Dumping hot water might harm it, for instance.

ACTIVITY 26 A change in the weather

Is the temperature always the same? What seasonal temperature changes were recorded? The conclusion of weather observations from Activity 4.

Science skills
organizing
inferring
communicating

Concepts
In temperate and subtropical regions there are annual seasonal temperature changes which may influence plant and animal biology and behavior.

Skills practiced
graphing

Time
1–2 periods

Mode of instruction
class discussion
independent student group work

Sample objectives
Students organize large temperature data sets and graphically display them.

Students predict future temperature data based on past trends.

Builds on
Activity 4
Activity 25

Teacher's information

Seasonal temperature changes and variations in rainfall occur in TEMPERATE (freezing weather occurs each year) and SUBTROPICAL (freezing occurs some years) climates. These variations are important to plants and animals whose entire life cycles may be keyed to these changes. Even in TROPICAL (no freezing temperatures) climates there are seasonal variations in temperature and rainfall that strongly influence the life cycles of plants and animals.

Introduction

Review the importance of temperature in aquatic ecosystems and as a variable in Activities 20, 22, 23 and 25. How do temperatures change seasonally where you live? What are the easiest ways to look at the data you have collected?

Action

Distribute the temperature and rainfall data charts recorded by the students or copied from a geographic atlas for your region. Each group might get one month of data. Let groups discuss how they might reduce these numbers to easy to read or interpret graphs. Ask questions about plotting each day versus weekly or monthly totals (rainfall) or averages (temperature). What gives the clearest picture? Have the students make a list of characteristics of a good graph. This develops a scoring rubric for their work. Give them graph paper and colored pencils and let them test their ideas.

Results and reflection

Lead a student discussion on the various techniques used and have the class select the ones they found the easiest to understand. One of the greatest challenges to scientists is to reduce piles of data to easily understood charts and graphs. Selecting to average data by week or month can be very useful. Use butcher paper or the board (paper can be left up and continued) to put together seasonal or annual temperature and rainfall graphs for your region. In general, bar graphs are used for rainfall. Temperature data may be displayed in line or bar graphs and maximum and minimum temperatures may be displayed on the same graph.

If daylength data are available and plotted, ask students to discuss correlations between temperature and day length. Does the increase in one lag behind the increase in the other? Can they speculate on why? It takes a long while to warm things up.

Materials

seasonal temperature and rainfall
   data from Activity 4 or annual
   temperature and rainfall records
   for your region
graph paper (see Recipes)
colored pens or pencils
butcher paper and markers or
   colored chalk and blackboard
optional
day length data

Preparation

Collect data from Activity 4 begun at the start to the school year or data from an atlas, newspaper weather information or Internet weather data base.

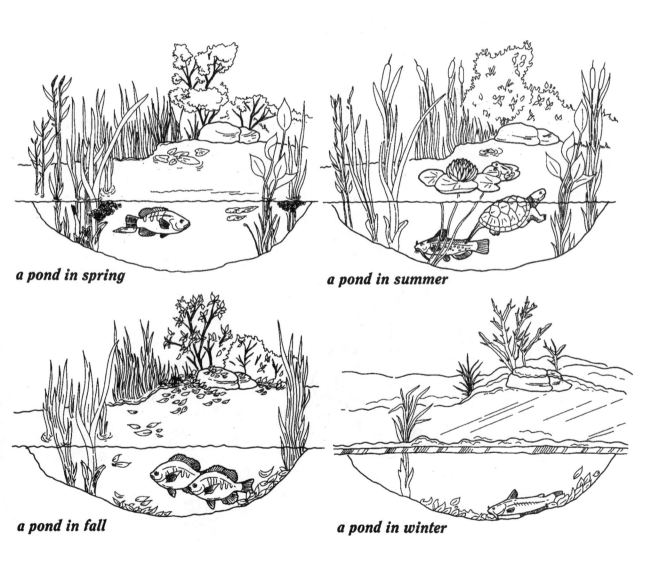

a pond in spring

a pond in summer

a pond in fall

a pond in winter

Outline

before class

1. copy graph paper

2. collect data

during class

1. distribute data

2. lead class discussion in its use and display

3. student groups experiment with graphs and displays

4. discuss and refine graphs

Conclusions

The physical characteristics of local environments vary through time. Temperature and rainfall are variations that are easily measured. Temperature varies as a function of daylength, with temperature increases and decreases lagging behind changes in daylength as the earth and water slowly warm and slowly cool.

Extension and applications

1. Select weather data from one or more regions that are geographically very different from your area. Have students plot and compare them with their own data. Can they speculate on how these different conditions might affect plants and animals that live in water?

2. Assign independent library research projects on a range of aquatic organisms. Ask students to describe the annual seasonal variations in biology and behavior of the animals or plants. For example: salt marsh grasses, beavers, phytoplankton, northern pike, diamondback turtles, frogs, water lilies, humpback whales. When do they grow? Reproduce? Rest or hibernate? Have the student products displayed as one 8.5 by 11 inch sheet and make a bulletin board divided by habitat. Can the students draw any conclusions from the displays about seasonal variations in life cycles?

ACTIVITY 27 The great anadromous fish game

What are some of the factors that determine the reproductive success of fish that live in the ocean and spawn in fresh water during the seasonal migrations? A board game that simulates the life history of blueback herring.

Teacher's information

This game is the original work of a classroom teacher, Martin Tillett of the Howard B. Owens Science Center, Prince George's County Public Schools. He developed it while working on weekends as an environmental educator at Herring Run Park. It has been modified by the National Aquarium in Baltimore education staff, in part based on the comments of teachers and fisheries biologists who have played it. Don Flescher of the National Marine Fisheries Service in Woods Hole, MA has used and reviewed it.

MIGRATION is the movement of animals from one area to another. Many species migrate annually during specific SEASONS of the year. In this game, blueback herring seasonally migrate from the open ocean through estuaries and into freshwater rivers and streams where they SPAWN (lay their eggs). The newly hatched young must then migrate back down the rivers to the sea. Fish that live in the ocean and spawn in rivers are said to be ANADROMOUS from the Greek for "running upward." Both the adults and the young face a number of hazards, some natural and some of human origin. As the students play this game, they will learn about these hazards. Bluebacks were once incredibly numerous along the Atlantic Coast of the United States and southern Canada, but are severely reduced in numbers today. These changes caused by humans are explained in this game.

The math involved in calculating the reduction in numbers caused by these hazards requires advance practice for some age levels. One teacher with fourth graders "borrowed" some fifth graders to help her teach her class how to use calculators to do the math. After playing the game, students may wish to experiment with "what if" questions by removing some of

Science skills
observing
organizing
communicating

Concepts
Many fish species migrate seasonally for reproduction.

While each fish is capable of producing large numbers of offspring, many factors reduce the chance that an individual fish will reproduce and that its offspring will survive.

Mortality in anadromous fish populations may be due to natural causes or may be caused by human actions.

Causes of mortality vary with the age and location of the fish.

Skills practiced
multiplying by fractions or decimals
subtraction
rounding off
using a calculator
graphing

Time
2–3 periods

Mode of instruction
independent group work

Sample objectives

Students describe the seasonal migration of an anadromous fish.

Students identify a variety of natural and human factors that affect the reproductive success of anadromous fish.

Students apply mathematical skills to the study of a biological problem.

Builds on

Activity 26

Materials

for each group (2–6 players)
1 game board
sets of hazard cards
1 die
storage box such as a shirt box
vocabulary and rules sheet
calculator if not doing math by hand
large sheet (18 x 24 inches) of
    construction paper
for each student
worksheet
colored marker for game
graph paper (see Recipes)
optional for class
video tape *The River Herring* (order from Photography by Michelson, P.O. Box 850093, Braintree, MA 02185; (617) 848-8870; price was $14.95 in 1996; included shipping; excellent; 7 minutes)

the hazard cards, such as suspending all fishing, to see what effects these changes have.

Introduction

Start by asking the class if they think animals always stay in the same place or do they move? If so, why? MIGRATION should come up in the discussion. Have the students list some animals that migrate and suggest reasons why they might do seasonal migrations—climatic changes affect food supply and reproductive potential. For example, humpback whales migrate to cold, rich northern waters to feed in summer and swim south to warmer water to calve during the winter. Canada geese migrate north each spring to breed in the northern U.S. and Canada and south each fall to winter feeding grounds in the southern regions of the United States.

What about fish? Here is a game in which the students get to be herring which migrate to the ocean to feed as adults and into rivers and creeks to SPAWN, releasing eggs which are fertilized outside the female's body. The fish in this game live along the eastern coast of the United States. Review the term ESTUARY as these fish will pass through an estuary both coming from and going to the spawning grounds. Have the students predict some of the hazards they are likely to encounter during their migration: fishing, predators, bad weather, lack of food, human barriers like dams, pollution.

You may also show the video tape *The River Herring—Bluebacks and Alewives* which may be ordered from Photography by Michelson, P.O. Box 850093, Braintree, MA 02185. A copy was $14.95 in 1996. This excellent tape was made by the Massachusetts Fish and Game Division. It sets up the game nicely by showing herring, and their migration, reproduction and development in about 7 minutes.

Action

Pass out game sets to groups of students and let them work independently to play the game and fill out their worksheets. Note: the students will be very upset to see how fast their baby herring population declines. Make sure they understand that if they each get only two herring back to the spawning ground, they will have replaced themselves and the population will remain stable.

Results and reflection

Compare the numbers of fish that successfully completed the annual migration. Have each student plot the number he/she finished with on the board on a graph so that everyone can see the variation in outcome. Discuss the role of chance in various outcomes. What was the average, range? On the whole, was the herring population growing or decreasing? In this game at what stages in their lives are herring most vulnerable?

Why might we want to have healthy herring populations? Were they important to other species? Were they important for humans? Make lists of the causes of mortality. Ask the students for suggestions about ways humans could modify laws or the environment to increase the number of herring.

Have the students discuss the nature of this game as a model of a real situation. How is it useful? Can they manipulate the model in ways that they could not change the real system in order to ask "what if" questions?

Conclusions

These are among the things which naturally reduce the size of anadromous fish populations:

- predation by a wide variety of animals
- limited food supplies
- changes in salinity and water level from unusual rainfall
- abnormal temperatures
- unusually severe storms
- parasites and diseases

These are some of the many ways humans have reduced herring populations:

- waste products released into the water by humans that are either directly toxic or kill by lowering oxygen level

Preparation

This game requires time in its initial construction. Once made, it can be taken off the shelf and used. As with anything you plan to use repeatedly, lamination makes it last longer. Xerox each kind of card (reproduction, stream to ocean, ocean, etc.) onto different colors of paper so they can be sorted easily. Be careful to get fronts and backs matched as they are in this book. You or your students might color to the illustrations on the game boards before lamination. Copy the worksheets. Copy the four parts to the game board and attach to poster sized construction paper. Before playing the game, you will need to review multiplying by fractions or decimals and rounding off with your students. For example, you have 75,000 fish and one quarter of them are eaten by a school of striped bass. If one quarter are eaten, three quarters or 0.75 survive. Multiply 75,000 by 0.75 to get 56,250 fish still alive. If math is being done by hand, the numbers may be rounded off to the nearest thousand to speed calculations.

Outline

before class

1. make games

2. practice the math

during class

1. discuss seasons and seasonal migration

2. use herring as an example

3. show herring video if you have it

4. pass out games and let the students play

5. record data as played

6. analyze group data

7. shared data with class

8. discuss; list causes of mortality

9. may change rules and replay

- accidental entry of toxic compounds with runoff such as pesticides
- sediment from runoff of farms or developments
- obstructions to migration such as dams
- over fishing by a wide variety of methods
- water heated by human activities

Even if humans are totally out of the picture, far more herring are spawned than will ever survive to reproduce. Each species of animal or plant is capable of producing more offspring than are needed to just replace the individuals already alive. This allows species to survive predation and recover from natural changes or disasters. It also means that when natural controls such as predators are removed, populations may explode in size.

Extension and applications

1. What would happen if human-caused fish deaths were reduced? Let the students choose one set of conditions to change, such as no longer allowing any fishing. Replace these cards with blank cards and see what happens to the numbers. Would they continue to increase forever? No, because eventually the predator populations would increase and/or competition for food supplies would occur. These would set new limits for the numbers of herring. We do know from historical records that when European people first arrived in North America, there were far more herring than there are today. Many of the problems we have created are not new. Dams for power and the location of industries along rivers began more than two hundred years ago.

2. Have students choose an aquatic or marine species that migrates. Research its life history in the library or on the Internet. Then make their own realistic game that is based on that migration. When the games are done, trade them around and let other students play them.

The great anadromous fish game rules

Read all the instructions first!

Goal

You are an adult female fish at the spawning ground. Your objective is to produce as many offspring that return to the spawning grounds as possible. You will play the game to see how many of the baby herring get to the ocean to grow up and then return to spawn. You only need two fish to make it back to replace your female in the population. The player with the most returning fish wins! But beware, there are many hazards lurking along the way.

How to play

1. Open the board. Shuffle each set of hazard cards and place them in the marked locations.

2. Select your marker and place it in the Spawning Grounds. This is your starting point as an adult female herring. Record 100,000 as the number of eggs you carry. Pick a reproduction card to find out how many of your fertilized eggs survive. Record this number. Your offspring will now swim through the Streams and Creeks and then into Rivers and the Estuary and out to the Ocean where they will feed and grow over several years. Survivors swim back into the Estuary, upriver and then upstream to spawn in the Creek where they were born.

3. Roll the die. The highest number starts first. The person to the left goes next. Play proceeds clockwise from that player.

4. Roll the die to find out how far you move your marker. If you land on a space that says to draw a card, do so and read it aloud. If you picked a

hazard card, you must record the change in number of fish in your school on your worksheet and the cause of mortality or death. For example, if you have 50,000 herring left and your card says that half of them were caught in nets, then you must reduce your herring to 25,000 fish and make note under human causes "fishing."

5. If you pick a hazard card that wiped out your entire school going TO the Ocean, record that and start completely over again at the beginning. If you are wiped out coming back to the Stream spawning grounds, you are finished.

6. Look up words that are new to you on the Vocabulary List.

7. Going to the Ocean, draw only cards that say TO Ocean. Once you arrive at the ocean, your fish will live there for several years before returning to spawn.

8. When mature, your fish will return in early spring to the Estuary from which they came and swim back up to the exact place where they were born, to produce a new generation.

9. The player who gets the most fish back to Spawning Grounds WINS, not the player who gets there first.

Vocabulary

anadromous: fish that live as adults in the ocean, but migrate back to the rivers or streams where they were born to reproduce; salmon and striped bass are good examples

bait: fish or other animal used to lure a larger animal into a trap or onto a hook

bluefish: a species of fish larger than herring which eat smaller fish; they follow prey into the estuary and are known for their feeding frenzies; are taken by humans for sport and food

commercial: things that are sold; activity done to earn money or wages

delicacies: expensive, rare things to eat

dolphin: a marine mammal in this game; erroneously called a porpoise

gill nets: nets which hang from floats in the water and catch fish that try to swim through the holes by tangling their gill covers

haul seine: large net pulled from shore by boat to circle fish school and then towed up on beach with catch; this method is not used much

kippers: herring that is salted and then smoked over a fire to preserve it

larvae: baby fish which look different than adult; hatch from eggs and change into the adult form as they grow

leech: a worm that sucks blood from its prey; a parasite

pesticide: a chemical used to kill insects or other small animals that humans regard as bad (pests)

plankton: tiny drifting plants and animals that live in water; eaten by fish

purse seine: large circular net used from boat to catch fish; draws shut at bottom

school: a large number of fish that swim together

seine: net with floats at top and weights at bottom used to encircle fish

snag hooks: hooks designed to catch on a fish's body

sport: an activity done for fun or excitement that requires skill

striped bass: large predatory fish; also anadromous; they enter estuaries in search of smaller fish and to spawn

trawl nets: large nets towed behind a boat that scoop fish up

tuna: very big predatory fish that live in the ocean

weir traps: pronounced "weer"; traps that funnel fish through a narrow space where it is easy to dip them from shore with hand-held nets

Reproduction

You got enough to eat and your spawning grounds were protected by students doing a stream restoration project. You have 100,000 offspring.

Reproduction

A big spring storm strikes right after you laid your eggs. It brings sediment into your small stream from fields where farmers plowed recently, preparing to plant corn. The mud kills most of your eggs and only 2,000 survive.

Reproduction

A late ice storm hits. Salt used to melt ice on the roads runs into the stream and kills most of your eggs. Only 5,000 survive.

Reproduction

The weather is good and only a few of your eggs die. 80,000 survive to begin migration.

Stream to Ocean

Small predatory fish in the stream eat 1/2 (50%) of your larval fish.

Reproduction

A big herd of cows is released into the pasture where your stream is. They walk through the stream, stirring up the bottom. 75,000 of your eggs survive.

Reproduction

You release your eggs on a Friday. That night at a chemical plant upstream someone opens a valve to dump toxic waste illegally. He hopes no one will be testing the stream on the weekend. Only 12,000 of your offspring survive.

Reproduction

A very cold early spring storm kills some of your eggs. 50,000 survive.

Reproduction

Your eggs are in a small, shallow stream when acid rain from air pollution falls. The soil in this stream does not neutralize the acidity and only 25,000 of your young survive.

Stream to Ocean

Pesticide runoff from a nearby farm poisons 1/4 (25%) of your young fish.

REPRODUCTION CARDS

REPRODUCTION CARDS

REPRODUCTION CARDS

REPRODUCTION CARDS

REPRODUCTION CARDS

REPRODUCTION CARDS

REPRODUCTION CARDS

REPRODUCTION CARDS

**STREAM AND CREEK
CARDS
To Ocean**

**STREAM AND CREEK
CARDS
To Ocean**

Stream to Ocean
Your school was killed before it hatched into larvae when mud from a new housing development smothered all (100%) of the fertilized eggs.

Stream to Ocean
A huge rainstorm kills 1/2 (50%) of your school as the tiny larvae are tumbled against rocks during flooding.

Stream to Ocean
Passing through toxic chemicals leaking into the stream from an illegal waste dump, 1/2 (50%) of your school dies.

Stream to Ocean
None of your larvae die. The spawning grounds are protected by laws which preserve their natural state. Advance one space.

Stream to Ocean
3/4 (75%) of your school dies after entering a section of the stream where industrial pollutants were dumped illegally.

Stream to Ocean
1/4 (25%) of your school is left stranded in shallow pools by a passing flood. They cannot get back to the creek and die.

Stream to Ocean
Chlorine entering the water from a small sewage treatment plant kills 1/2 (50%) of your school.

Stream to Ocean
None of your school dies. The stream where they hatched was cleaned up by school children and protected as park.

Stream to Ocean
1/2 (50%) of your school dies because the food the larvae need is not abundant this year due to weather.

Stream to Ocean
Swimming into an area of very hot water where the stream is being used to cool heavy equipment, 3/4 (75%) of your school dies

**STREAM AND CREEK
CARDS
To Ocean**

**STREAM AND CREEK
CARDS
To Ocean**

**STREAM AND CREEK
CARDS
To Ocean**

**STREAM AND CREEK
CARDS
To Ocean**

**STREAM AND CREEK
CARDS
To Ocean**

**STREAM AND CREEK
CARDS
To Ocean**

**STREAM AND CREEK
CARDS
To Ocean**

**STREAM AND CREEK
CARDS
To Ocean**

**STREAM AND CREEK
CARDS
To Ocean**

**STREAM AND CREEK
CARDS
To Ocean**

River and Estuary to Ocean

After swimming through an area polluted with industrial wastes that are toxic to young herring. 3/4 (75%) of your school dies.

River and Estuary to Ocean

None of your school dies. It has a safe passage toward the sea. Advance one space.

River and Estuary to Ocean

Your school swims into an area where an algal bloom has died. The rotting algae used all the oxygen in the water. 1/2 (50%) of your school dies from lack of oxygen.

River and Estuary to Ocean

1/4 (25%) of your school dies after being attacked by brackish water fish leeches. These parasites weakened the young herring.

River and Estuary to Ocean

1/4 (25%) of your school is eaten by eels, wading birds, water snakes and other natural predators on small fish.

River and Estuary to Ocean

An unusually dry summer caused salinity changes. Lack of food starves 1/4 (25%) of your school.

River and Estuary to Ocean

None of your school dies. It avoided predators, was not exposed to toxic wastes because laws have helped control wastes, and found normal food supplies and weather.

River and Estuary to Ocean

1/2 (50%) of your school dies after swimming through water polluted with manure runoff from farms. Decaying manure used all the oxygen in the water.

River and Estuary to Ocean

Swimming into the cooling water intake pipe of a steel mill, 3/4 (75%) of your school dies.

River and Estuary to Ocean

Hungry gulls and other birds eat 1/2 (50%) of your school when it is caught in a shallow area of the river caused by unusually dry weather.

RIVER AND ESTUARY CARDS
To Ocean

RIVER AND ESTUARY CARDS
To Ocean

RIVER AND ESTUARY CARDS
To Ocean

RIVER AND ESTUARY CARDS
To Ocean

RIVER AND ESTUARY CARDS
To Ocean

RIVER AND ESTUARY CARDS
To Ocean

RIVER AND ESTUARY CARDS
To Ocean

RIVER AND ESTUARY CARDS
To Ocean

RIVER AND ESTUARY CARDS
To Ocean

RIVER AND ESTUARY CARDS
To Ocean

River and Estuary to the Ocean
A large school of hungry bluefish eat 1/2 (50%) of your school. The rest escape as sport fishermen scare the bluefish away while trying to catch them.

River and Estuary to Ocean
None of your school dies. Advance one space.

Ocean
Commercial fishermen using gill nets catch 1/2 (50%) of your school. They are used to make plant fertilizer.

Ocean
None of your school dies as the weather has been perfect. Advance one space.

Ocean
Commercial fishermen using trawl nets catch 1/4 (25%) of your school. They are ground into fish meal for chicken feed.

Ocean
Hungry dolphins surround and herd the school. They eat 1/2 (50%) of your fish.

Ocean
Commercial fishermen using a haul seine catch 1/4 (25%) of your school. They are sold for bait for crab and lobster traps and for fish bait.

Ocean
Commercial fishermen using gill nets catch 1/2 (50%) of your school. The fish are made into kippers and pickled herring.

Ocean
None of your school dies. It found enough food, was not caught by predators, and encountered normal weather.

Ocean
Commercial fishermen using gill nets trap 3/4 (75%) of your school. The roe (eggs) are processed and sold as delicacies in Japan.

RIVER AND ESTUARY CARDS
To Ocean

RIVER AND ESTUARY CARDS
To Ocean

OCEAN CARDS

OCEAN CARDS

OCEAN CARDS

OCEAN CARDS

OCEAN CARDS

OCEAN CARDS

OCEAN CARDS

OCEAN CARDS

Ocean

Commercial fisherman using a purse seine encircle 3/4 (75%) of your school. The fish are processed into fish oil for use in paint, medicines and cosmetics.

Ocean

Commercial fishermen take 3/4 (75%) of your school for use as fish bait.

Ocean

A school of striped bass eats 1/4 (25%) of your school.

Ocean

1/4 (25%) of your school is eaten by striped bass.

Ocean

A school of hungry bluefish and a flock of seabirds eats 1/4 (25%) of your school.

Ocean

None of your school dies. They escaped predators and found plenty of plankton to eat.

Ocean

1/4 (25%) of your school is eaten by a large flock of hungry gulls which caught them at the surface where the school went to escape from a bluefish school.

Ocean

Commercial fishermen seeking immature fish for bite-sized snacks catch 1/2 (50%) of your school.

Ocean

A hungry school of bluefish eat 1/2 (50%) of your school.

Ocean

3/4 (75%) of your school dies due to lack of sufficient plankton, the food on which you depend.

OCEAN CARDS

OCEAN CARDS

OCEAN CARDS

OCEAN CARDS

OCEAN CARDS

OCEAN CARDS

OCEAN CARDS

OCEAN CARDS

OCEAN CARDS

OCEAN CARDS

Ocean

None of your school dies. They escaped predators and got plenty of plankton to eat.

Ocean

1/4 (25%) of your school is eaten by a large school of tuna.

Estuary and River to Spawning Grounds

Your school is trapped in shallow water. A large flock of seagulls eats 1/4 (25%).

Estuary and River to Spawning Grounds

Commercial fishermen using haul seines catch 1/2 (50%) of your school. The fish become smoked herring.

Estuary and River to Spawning Grounds

1/4 (25%) of your school dies as it gets pulled into the cooling water intake of a huge electric power plant that has no fish screen.

Estuary and River to Spawning Grounds

Commercial fishermen sell their catch to an aquarium for seal food. 1/2 (50%) of your school is lost.

Estuary and River to Spawning Grounds

A school of hungry striped bass eats 1/4 (25%) of your school.

Estuary and River to Spawning Grounds

1/2 (50%) of your school dies when part of the school takes a fork in the river that leads to a dam with no way around.

Estuary and River to Spawning Grounds

Commercial fishermen use gill nets staked out along the shore to catch 1/2 (50%) of your school. The fish are made into salt herring.

Estuary and River to Spawning Grounds

Sport fishermen line the banks of the narrow river channel. They hook 1/4 (25%) of your school using snag hooks. The fish will be smoked or used as bait.

OCEAN CARDS

OCEAN CARDS

ESTUARY AND RIVER CARDS
Return To Spawning
Grounds

ESTUARY AND RIVER CARDS
Return To Spawning
Grounds

ESTUARY AND RIVER CARDS
Return To Spawning
Grounds

ESTUARY AND RIVER CARDS
Return To Spawning
Grounds

ESTUARY AND RIVER CARDS
Return To Spawning
Grounds

ESTUARY AND RIVER CARDS
Return To Spawning
Grounds

ESTUARY AND RIVER CARDS
Return To Spawning
Grounds

ESTUARY AND RIVER CARDS
Return To Spawning
Grounds

Estuary and River to Spawning Grounds
An early spring storm blows 1/4 (25%) of your school ashore when they are in shallow water.

Stream to Spawning Grounds
Your stream was made into a ditch as a flood control project since your school was born and left it. Most spawning sites are destroyed. All (100%) of your school dies without spawning.

Estuary and River to Spawning Grounds
A school of hungry bluefish eat 1/4 (25%) of your school.

Estuary and River to Spawning Grounds
Commercial fishermen using weir traps and dip nets in narrow sections of the river trap 1/2 (50%) of your school. The fish are sold fresh to a seafood market.

Stream to Spawning Grounds
Improper farming methods have choked the stream with mud. 1/2 (50%) of your school is smothered with mud in their gills.

Stream to Spawning Grounds
Your school enters a stream with pesticides from farm runoff. 1/2 (50%) of your school is poisoned.

Stream to Spawning Grounds
1/4 (25%) of your school is eaten by seagulls which have followed your schools upstream into shallow water.

Stream to Spawning Grounds
There has been no rain or snow this winter or early spring. 1/2 (50%) of your school dies when it strands in low water.

Estuary and River to Spawning Grounds
None of your school dies. Conditions for the trip through the estuary and up the river were excellent. Advance one space.

Stream to Spawning Grounds
The forest along the stream was cut. Stumps and logs form dams which your fish cannot cross. All (100%) of your school dies.

ESTUARY AND RIVER CARDS
Return To Spawning
Grounds

ESTUARY AND RIVER CARDS
Return To Spawning
Grounds

ESTUARY AND RIVER CARDS
Return To Spawning
Grounds

ESTUARY AND RIVER CARDS
Return To Spawning
Grounds

STREAM AND CREEK CARDS
Return to spawning
grounds

STREAM AND CREEK CARDS
Return to spawning
grounds

STREAM AND CREEK CARDS
Return to spawning
grounds

STREAM AND CREEK CARDS
Return to spawning
grounds

STREAM AND CREEK CARDS
Return to spawning
grounds

STREAM AND CREEK CARDS
Return to spawning
grounds

Stream to Spawning Grounds
None of your school dies. The stream you enter is protected as part of a park. Dams were removed, sediment is controlled and fishing is limited. Advance one space.

Stream to Spawning Grounds
Your school enters a stream next to an old toxic waste dump. Chemicals leaking into the water poison 1/2 (50%) of your school.

Stream to Spawning Grounds
Predators such as raccoons and river otters eat 1/4 (25%) of your school as it swims upstream.

Stream to Spawning Grounds
Your home stream has been blocked temporarily by a highway project. All (100%) of your school is unable to reproduce. You are wiped out.

Stream to Spawning Grounds
Local fishermen with nets and traps in the shallow, narrow creek catch 1/2 (50%) of your school. The fish are used for food.

Stream to Spawning Grounds
Your school enters a stream with an old, leaky sewer pipe running next to it. 1/4 (25%) of your school dies from lack of oxygen which is used as the sewage decays.

STREAM AND CREEK CARDS
Return to spawning grounds

STREAM AND CREEK CARDS
Return to spawning grounds

STREAM AND CREEK CARDS
Return to spawning grounds

STREAM AND CREEK CARDS
Return to spawning grounds

STREAM AND CREEK CARDS
Return to spawning grounds

STREAM AND CREEK CARDS
Return to spawning grounds

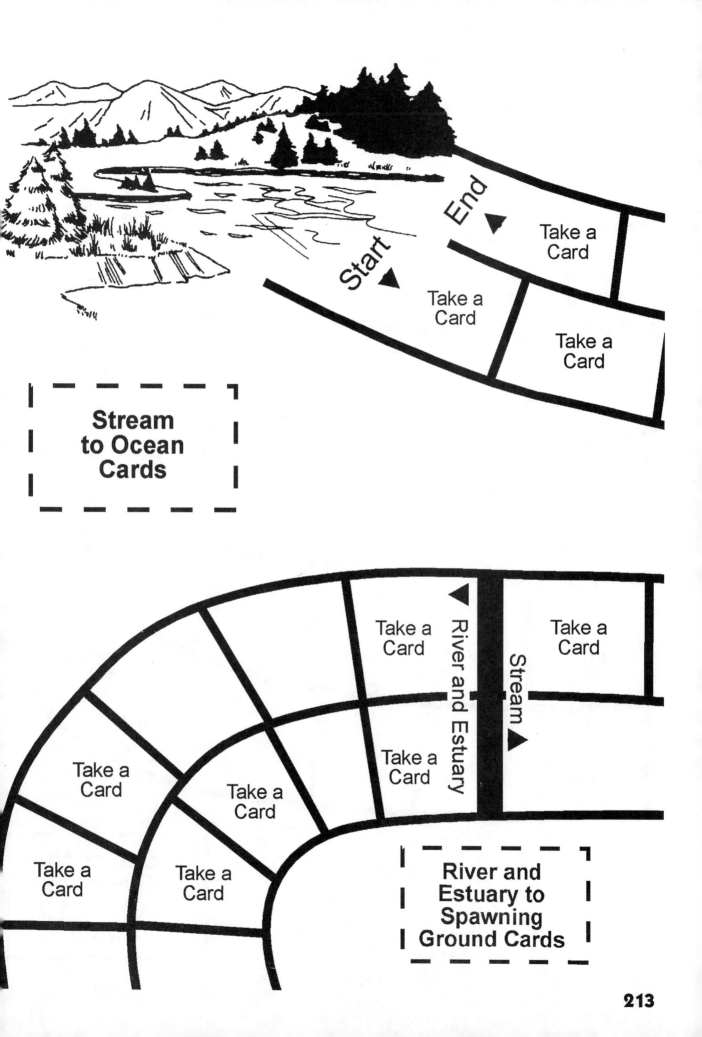

Stream
to Ocean
Cards

Start

End

Take a Card

Take a Card

Take a Card

Take a Card

River and Estuary

Stream

Take a Card

Take a Card

Take a Card

Take a Card

Take a Card

Take a Card

Take a Card

River and
Estuary to
Spawning
Ground Cards

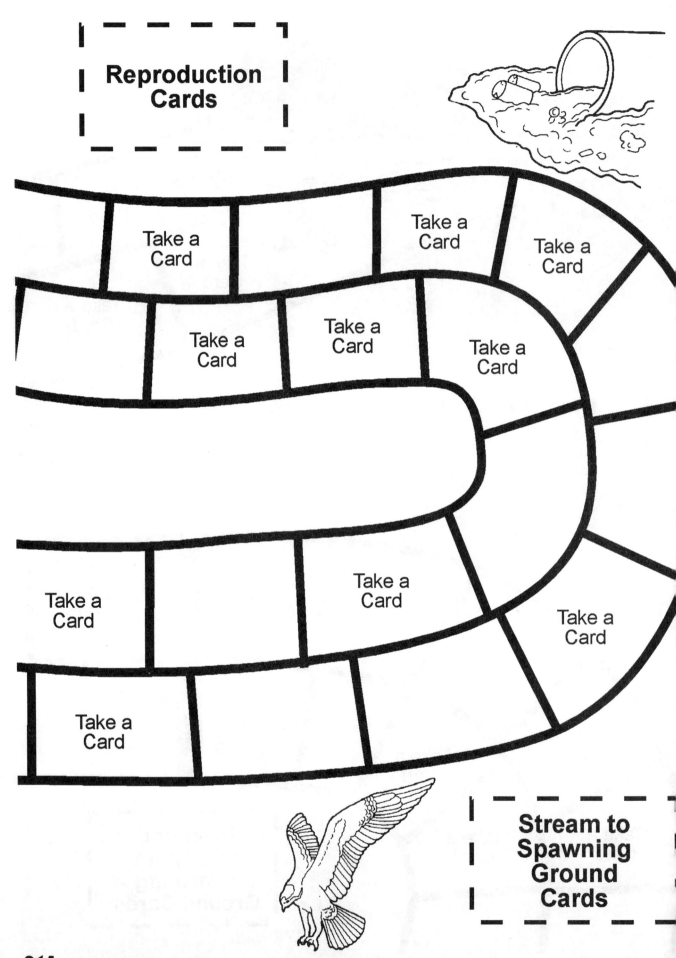

Reproduction
Cards

Take a
Card

Take a
Card

Take a
Card

Take a
Card

Take a
Card

Take a
Card

Take a
Card

Take a
Card

Take a
Card

Take a
Card

Stream to
Spawning
Ground
Cards

214

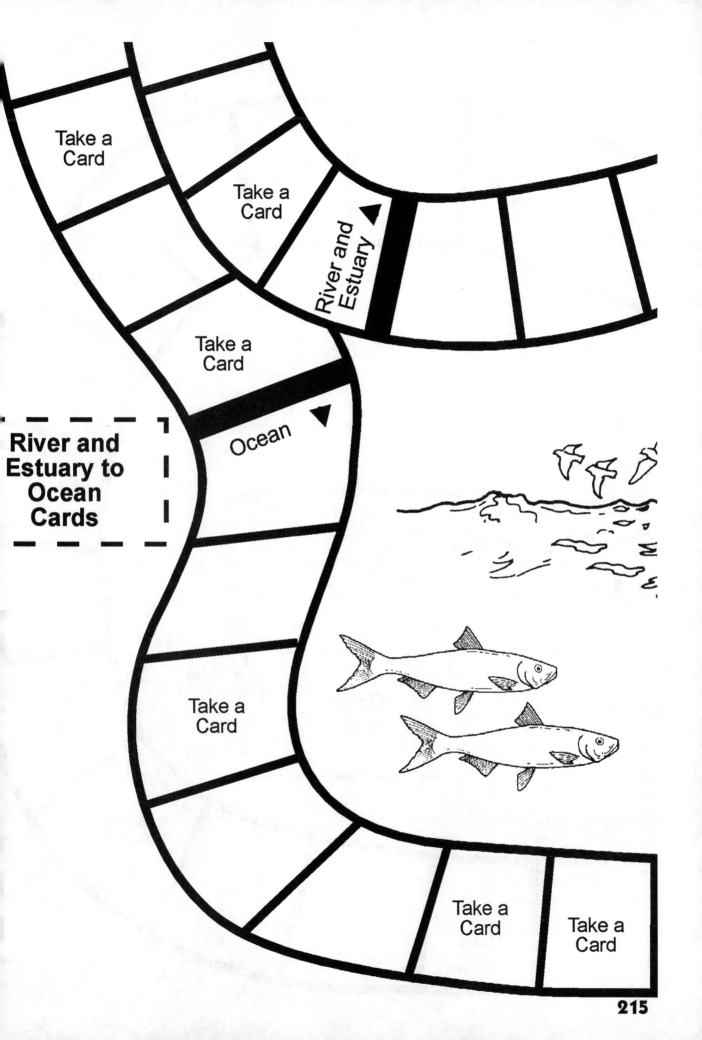

Take a
Card

Take a
Card

River and
Estuary ▲

Take a
Card

Ocean ▼

**River and
Estuary to
Ocean
Cards**

Take a
Card

Take a
Card

Take a
Card

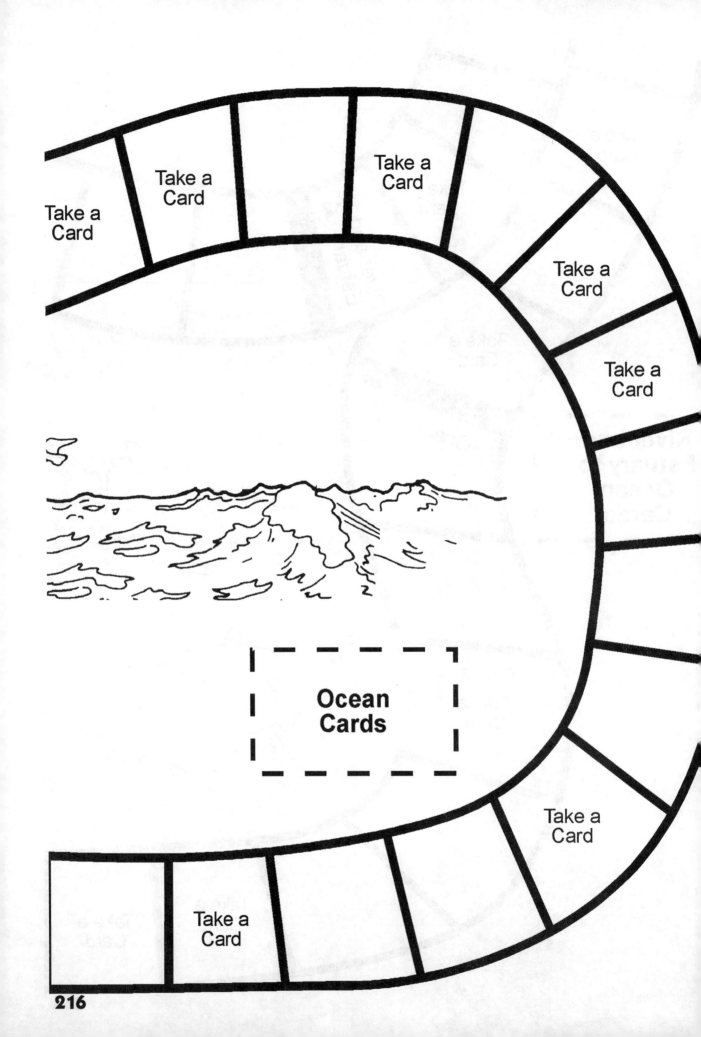

Take a Card

Take a Card

Take a Card

Take a Card

Take a Card

Take a Card

Ocean Cards

Take a Card

Take a Card

Activity 27
The great anadromous fish game

Name _____

1. You are a female herring at the spawning ground carrying 100,000 eggs. You release your eggs and they are fertilized in the water. Pick a Reproduction card to find out how many fertilized eggs survived. Record them here: _____ offspring. This is the size of your school at the beginning of the game.

2. Play the game and record the numbers of your school and the causes of their death (mortality) on the Fish School Data Table.

3. After you finish the game enter the numbers that remained at the end of each section of the migration for each player in your group.

person	100,000 eggs in female	# of eggs that survive	# young at end of stream	# young at end of estuary	# adults at end of ocean	# adults at end of estuary	# adults at spawning ground
1							
2							
3							
4							
5							
6							
total							
average							

4. Each of you played the same game. Explain why you got different numbers of fish at the end.

5. Graphically display what happens to herring at different stages by showing the average school size at each stage of the herring life cycle shown in question 3. Use a piece of graph paper. Work together in your group to develop ideas about a good graph to display your group's game data from question 3. Then each of you should make your own graph of your group's results. Compare and refine your work. Attach it to this paper.

6. Describe the changes in numbers of herrings as they grow from fertilized eggs to adults.

7. What percent of your group's eggs grew into fish that returned to spawn?

8. If you were a fisheries manager, list three specific changes you would make to increase the number of herring.

9. Assume half the returning fish are female. How many of your school have to return to replace their mother?

10. Based on the returning fish from your group, could your group's population of fish increase in the next generation? Explain your answer.

Fish school data table (one for each player)

Direction	Location	number before hazard	fraction or % that die	kind of natural hazard	kind of human hazard	number after hazard (to next line)
# of fertilized eggs that survived						
To Ocean	Stream and Creek					
	Stream and Creek					
	Stream and Creek					
	Stream and Creek					
# of offspring at end of stream						
	Estuary and River					
	Estuary and River					
	Estuary and River					
	Estuary and River					
# of young fish at end of estuary						
Ocean	Ocean					
	Ocean					
	Ocean					
	Ocean					
	Ocean					
	Ocean					
# of adult fish at end of ocean						
Return to Spawn	Estuary and River					
	Estuary and River					
	Estuary and River					
	Estuary and River					
# of adult fish at end of river and estuary						
	Stream and Creek					
	Stream and Creek					
	Stream and Creek					
	Stream and Creek					
# of adult fish that reached spawning grounds						

Activity 27
The great anadromous fish game

Name __*possible answers*__

1. You are a female herring at the spawning ground carrying 100,000 eggs. You release your eggs and they are fertilized in the water. Pick a Reproduction card to find out how many fertilized eggs survived. Record them here: __50,000__ offspring. This is the size of your school at the beginning of the game.

2. Play the game and record the numbers of your school and the causes of their death (mortality) on the Fish school data table.

3. After you finish the game enter the numbers that remained at the end of each section of the migration for each player in your group.

person	100,000 eggs in female	# of eggs that survive	# young at end of stream	# young at end of estuary	# adults at end of ocean	# adults at end of estuary	# adults at spawning ground
1	100,000	50,000	3,125	1,650	39	39	9
2	100,000	2,000	187	70	18	7	2
3	100,000	80,000	1,875	132	19	7	0
4							
5							
6							
total							
average	100,000	44,000	1,729	617	25	18	4

4. Each of you played the same game. Explain why you got different numbers of fish at the end.

   What happened was due to the roll of the dice. It was chance which of us picked which card.

5. Graphically display what happens to herring at different stages by showing the average school size at each stage of the herring life cycle shown in question 3. Use a piece of graph paper. Work together in your group to develop ideas about a good graph to display your group's game data from question 3. Then each of you should make your own graph of your group's results. Compare and refine your work. Attach it to this paper.

6. Describe the changes in numbers of herrings as they grow from fertilized eggs to adults.

The numbers went down <u>real</u> fast and then kind of leveled out. Babies die most.

7. What percent of your group's eggs grew into fish that returned to spawn?

0.004% $\qquad \dfrac{4}{100,000} = 0.00004 \times 100\% = 0.004\%$

8. If you were a fisheries manager, list three specific changes you would make to increase the number of herring.

1. Remove all barriers to fish swimming back up stream.

2. Reduce fishing pressure.

3. Clean up toxic pollution.

9. Assume half the returning fish are female. How many of your school have to return to replace their mother?

1 female and 1 male or 2 fish

10. Based on the returning fish from your group, could your group's population of fish increase in the next generation? Explain your answer.

We only needed 2 fish on average to return but 4 made it back so our population could actually grow.

Fish school data table (one for each player)

Direction	Location	number before hazard	fraction or % that die	kind of natural hazard	kind of human hazard	number after hazard (to next line)
# of fertilized eggs that survived		50,000				
To Ocean	Stream and Creek	50,000	75%		pollution	12,500
	Stream and Creek	12,500	50%	weather		6,250
	Stream and Creek	6,250	50%	predation		3,125
	Stream and Creek					
# of offspring at end of stream		3,125				
	Estuary and River	3,125	50%		fishing	1,650
	Estuary and River					
	Estuary and River					
	Estuary and River					
# of young fish at end of estuary		1,650				
Ocean	Ocean	1,650	75%		fishing	412
	Ocean	412	0%			309
	Ocean	412	25%	predation		309
	Ocean	309	50%		fishing	155
	Ocean	155	75%		fishing	39
	Ocean	39				
# of adult fish at end of ocean		39				
Return to Spawn	Estuary and River	39	0%			
	Estuary and River	39				
	Estuary and River					
	Estuary and River					
# of adult fish at end of river and estuary		39				
	Stream and Creek	39	50%	predation		19
	Stream and Creek	19	50%		pollution	9
	Stream and Creek	9				
	Stream and Creek					
# of adult fish that reached spawning grounds		9				

ACTIVITY 28 Volcano!

Temperature and dissolved oxygen: a scenario in which students apply information and skills from a previous exercise to solve a problem.

Teacher's information

This is a made up construct (that water temperature rises with increasing volcanic activity), but credible. Students need to apply both their testing skills and their knowledge to solve the problem posed here. You can check their testing accuracy as well as their thinking with this assessment. The individual work at the end gives fast individuals something to do while the other groups are finishing up.

Introduction

Read the challenge at the top of the worksheet. Pass out the worksheets and materials and let them go.

Action

Students work independently, following the instructions on their sheets. You may decide how much help you want to give. If you answer a question for one group or pair, be sure to state the question and your answer for the entire class to avoid giving unfair advantage.

Science skills

measuring
organizing
inferring
communicating

Concepts

Information learned in one context may be applied to others.

Skills practiced

measuring dissolved oxygen
graphing

Time

1–2 periods

Mode of instruction

independent student pair and group work

Sample objectives

Students accurately test water for dissolved oxygen.

Students work cooperatively to pool and organize data.

Students make inferences and predictions based on data.

Students apply information to new situations and explain how they did so.

Builds on

Activity 23

Materials

for class

4 dissolved oxygen test kits

2 sets of 5–8 water samples in 1/2 pt
  canning jars (see Preparation)

for each pair (or individual)

dissolved oxygen sample bottle A
  (see Recipes)

dissolved oxygen titrator or syringe
  (see Recipes)

dissolved oxygen vial B (see Recipes)

for each student

worksheet

Preparation

Make water samples ahead of time
that represent "samples" taken from
the supposed volcanic pools. You
need two sets. If you have enough
materials and a class size of 16 or
fewer, each student can work alone
testing the samples. If you have more
than 16 students, they can work in
pairs. These are the instructions for
preparing 2 sets of 7 samples each
for 28 students to use in pairs. If you
had enough material, you could
make 3 sets for a large class or have
students work in groups of 3.

Line up two rows of 1/2 pt canning
jars, two deep and seven across.
Label the jars in reverse order. The
back row is 1a–7a reading from left
to right. The front row is 7b to 1b
reading from left to right. So 1a is in
behind 7b, 2a with 6b etc. Boil water
in a tea kettle for 15 minutes at a low
boil. You are going to make a series
with a continuous variation of the
proportion of cold tap water and
boiled water in each jar. Fill
according to this formula:

Results and reflection

The two sets of students should come to opposite
conclusions. One group (b) should say the pool is
cooling and the other (a), that it is heating up. Let
students compare means of presenting data, graphs
and logic. You may wish to discuss why the two
groups came to different conclusions. Students
working on the "a" samples should conclude that the
volcano is going to blow up while the "b" group had
water that was increasing in dissolved oxygen, and
hence cooling, indicating that the volcano was becoming quiet.

Conclusions

Collect the papers, read all of the answers, including
the individual work, and make comments and return
them for student portfolios.

	fill both jars heading each column with the combination of cold and hot water shown below those two numbers						
	1a and 7b	2a and 6b	3a and 5b	4a and 4b	5a and 3b	6a and 2b	7a and 1b
hot water	none	1/6 cup	1/3 cup	1/2 cup	2/3 cup	5/6 cup	all
cold water	all	5/6 cup	2/3 cup	1/2 cup	1/3 cup	1/6 cup	none

Top each jar off with a little of the boiled water so it is absolutely full and seal with a canning jar lid. Rather than measuring, it is possible to make the jars into a graded series by eyeballing them. Use cold water first to make them look like a line of glasses that you can play a tune on by tapping with a spoon. Then finish filling each up with boiling water and seal. The reason for this reverse order numbering is that the two groups are going to get different results. It makes for great confusion if the groups try to cheat by using each other's data!

Outline

before class
1. prepare water samples

2. copy worksheets

during class
1. read the prompt at the beginning of the worksheet

2. let them work independently

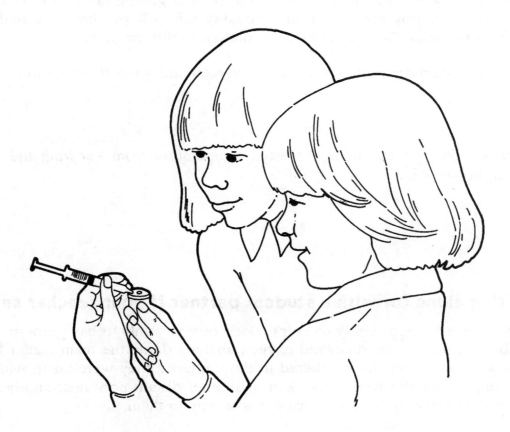

Activity 28
Volcano!

Name _____

Challenge: Is the volcano likely to erupt or not? Give the evidence for your answer. If you get ahead of others, turn to Question 5 at the end and work on it until the others catch up.

1. Working alone

Read this:

You and your partner are members of a team of geologists that has been working for several weeks near a volcano. You have been studying a pool fed by hot springs on the side of the volcano. You have been taking and storing water samples every other day. You sealed them for future chemical and dissolved oxygen testing.

Yesterday there was an earthquake which may mean the volcano is becoming more active. Your thermometers and the computer that stored your temperature data were destroyed in the earthquake. Fortunately, the dissolved oxygen test kits and the stored water samples survived. Everyone is wondering if the earthquake is a sign that the volcano is about to blow.

One team member suggests that temperature changes in the pool might be a way to tell if the volcano is becoming more active. If the pool is getting hotter, the volcano may blow! If it is not getting hotter, you are probably safe. All you have are sealed water samples from the pool. Can you think of a solution to this problem?

a. State the relationship between dissolved oxygen and water temperature:

b. Describe how you can infer the water temperature from knowing the dissolved oxygen in the pool.

2. Working alone (or with a student partner if your teacher says to)

The volcano may be threatening to erupt. None of you wants to hang around too long! To save time in getting the dissolved oxygen analysis done, the team leader has given you each a sample. They are numbered in the sequence they were taken with 1 being the first sample. Samples were also taken in another distant pool near another volcano which is also threatening. A second team is working on them.

a. Your sample code is _____

b. Determine the dissolved oxygen concentration of your sample. You may use the instructions for the dissolved oxygen test kit. If you are waiting for test kit materials or finish early, turn to the last page and work alone to solve the problems described there.

c. Record the dissolved oxygen in your sample _____

3. Sharing with your team

Assemble the members of your team (team a or team b) and work together on section 3.

a. Your samples are organized in a numbered sequence with the lowest number taken first. Arrange your group data in an appropriate table or chart here.

4. Working alone again

a. Your samples were taken at even time intervals. Make a graph displaying the relationship between time and temperature in your spring. (You do not need to know how long the intervals were to do this nor do you need to know the exact temperature.)

b. Given what you know about the relationship between water temperature and dissolved oxygen, what do you think is happening to water temperature in your spring?

c. If water temperature in pools fed by springs on a volcano rises as the volcano nears eruption, should you fear for your life if you stay on the volcano? Support your statement with data.

d. Describe how you used information and skills you learned in earlier activities to make your decision about the volcano.

5. Working entirely alone, complete this section.

You are a fish like a trout that needs high levels of dissolved oxygen. It is currently summer. You live in a temperate region where there are big lakes and small ponds, some surrounded by forests and some by farms. Describe the habitat you would prefer from among all these choices and give as much evidence as possible for why you would prefer it over the other options.

Activity 28
Volcano!

Name __possible answers__

Challenge: Is the volcano likely to erupt or not? Give the evidence for your answer. If you get ahead of others, turn to Question 5 at the end and work on it until the others catch up.

1. Working alone

Read this:

You and your partner are members of a team of geologists that has been working for several weeks near a volcano. You have been studying a pool feed by hot springs on the side of the volcano. You have been taking and storing water samples every other day. You sealed them for future chemical and dissolved oxygen testing.

Yesterday there was an earthquake which may mean the volcano is becoming more active. Your thermometers and the computer that stored your temperature data were destroyed in the earthquake. Fortunately, the dissolved oxygen test kits and the stored water samples survived. Everyone is wondering if the earthquake is a sign that the volcano is about to blow.

One team member suggests that temperature changes in the pool might be a way to tell if the volcano is becoming more active. If the pool is getting hotter, the volcano may blow! If it is not getting hotter, you are probably safe. All you have are sealed water samples from the pool. Can you think of a solution to this problem?

a. State the relationship between dissolved oxygen and water temperature:

> **We proved that as water temperature increases, dissolved oxygen decreases.**

b. Describe how you can infer the water temperature from knowing the dissolved oxygen in the pool.

> **If I can figure out dissolved oxygen over time and I know the relationship of dissolved oxygen to temperature, I can infer temperature.**

2. Working alone (or with a student partner if your teacher says to)

The volcano may be threatening to erupt. None of you wants to hang around too long! To save time in getting the dissolved oxygen analysis done, the team leader has given you each a sample. They are numbered in the sequence they were taken with 1 being the first sample. Samples were also taken in another distant pool near another volcano which is also threatening. A second team is working on them.

a. Your sample code is ___5a___

b. Determine the dissolved oxygen concentration of your sample. You may use the instructions for the dissolved oxygen test kit. If you are waiting for test kit materials or finish early, turn to the last page and work alone to solve the problems described there.

c. Record the dissolved oxygen in your sample ___3.8 ppm___

3. Sharing with your team

Assemble the members of your team (team a or team b) and work together on section 3.

a. Your samples are organized in a numbered sequence with the lowest number taken first. Arrange your group data in an appropriate table or chart here.

sample no.	dissolved oxygen ppm
1a	9.4
2a	7.9
3a	6.6
4a	5.2
5a	3.8
6a	2.4
7a	1.0

4. Working alone again

a. Your samples were taken at even time intervals. Make a graph displaying the relationship between time and temperature in your spring. (You do not need to know how long the intervals were to do this nor do you need to know the exact temperature.)

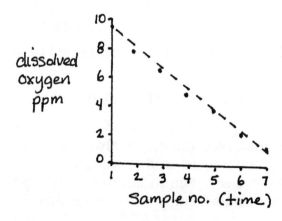

b. Given what you know about the relationship between water temperature and dissolved oxygen, what do you think is happening to water temperature in your spring?

Our spring is heating up because the dissolved oxygen is going DOWN!

c. If water temperature in pools fed by springs on a volcano rises as the volcano nears eruption, should you fear for your life if you stay on the volcano? Support your statement with data.

Yes, we should. Our data show the spring is heating up! That means the volcano is going to blow.

d. Describe how you used information and skills you learned in earlier activities to make your decision about the volcano.

We learned how to test for dissolved oxygen and are really good at it so we know we did it right. We did an experiment that showed the relationship between oxygen and water temperature.

5. Working entirely alone, complete this section.

You are a fish like a trout that needs high levels of dissolved oxygen. It is currently summer. You live in a temperate region where there are big lakes and small ponds, some surrounded by forests and some by farms. Describe the habitat you would prefer from among all these choices and give as much evidence as possible for why you would prefer it over the other options.

I need high dissolved oxygen so I want cold water which has more. That means I want deep water where it is cold in summer and a big body of water that has not heated up. That means I want to live in deep water in a big lake. I would also like clean, clear water so I would like forests with no sediment and no fertilizer.

SECTION IV Moving or staying put: maintaining position within aquatic habitats

Teacher's information

Surface tension

The physical characteristics of water have an effect on the nature of aquatic habitats and the kinds of places within those habitats where animals and plants may live. Water MOLECULES are the basic building blocks of this unique substance. Each molecule of water is composed of one ATOM of oxygen and two atoms of hydrogen. Due to the way in which these atoms are attached to each other, each water molecule has one slightly negative end and two slightly positive ends. The negative and positive parts of different molecules are attracted to each other when water molecules meet. This attraction is responsible for many of the characteristics of water. The weak positive and negative charges help make water the universal solvent studied in Section II.

The tendency of water molecules to stick to each other (COHESION) is also important in this section. The cohesive property of water is most obvious at the surface. The top layer of water molecules forms a strong elastic film or "skin" under the influence of SURFACE TENSION, the tendency of the membrane to contract to the minimum area. Many animals and plants live directly on the surface of bodies of water. Some of them float and are lighter than water like duckweed, a tiny green plant which floats in large clusters on the surface of ponds with its rootlets dangling in the water. Others are heavier than water and do not float. They rest on and are supported by the surface. Some kinds of beetles and bugs "walk on water" in search of prey.

The water strider, a familiar pond bug, has special hairs on its first and third pairs of legs which form dimples on the water surface. The strider's second pair of legs actually penetrate the surface and work like oars to propel the insect over the surface. There is a marine species of water strider as well. Other kinds of insects, like mosquito larvae, hang upside down from the surface film and poke a breathing tube up through it. Whirligig beetles are so well-adapted for life on the surface of ponds that each eye has two halves: the upper half can see above the surface while the lower half simultaneously views the underwater world!

The addition of soap to water interrupts surface tension by breaking the weak attraction among the water molecules. This explains why we add soap to our laundry—it allows the clothes to get wet. Some insects that might otherwise be preyed upon by water striders are able to release small amounts of detergent into the water. This disrupts surface tension and causes the strider to sink before it grabs its prey!

Density and buoyancy

Another important characteristic of water is its DENSITY. Some animals and plants are lighter than water of a volume equal to their own. These organisms float at the surface. The Portuguese man-of-war is a familiar marine invertebrate which floats. A very common misuse of the term "float" is in reference to plankton which do not float at the surface, but rather drift under water. Organisms that float are, by definition, those that live at the surface and are lighter than water.

Other creatures are just barely heavier than water and sink very slowly. This is a problem for organisms that need to stay near the surface. PHYTOPLANKTON, the tiny algae that form the basis of the food chain in many bodies of water, are heavier than water. If they were not, they would all float in a single crowded layer at the surface. Different kinds have different tactics to achieve the same end: to sink very slowly, staying as long as possible in the lighted zone. Some store food (energy) as oils rather than as the starch typical of higher green plants. The oil helps keep their density near that of the warmer surface waters. Another tactic is to have lots of long projections on their surface. These create drag, slowing their sinking. Also, many organisms we call plants that are phytoplankton have tiny "hairs" called FLAGELLA which may help them move when waved. Flagella are usually associated with animal cells in most people's minds, but many "plant" cells are also capable of movement. Yet another way to sink slowly is found among the diatoms—small flattened phytoplankton. As they sink, their shape causes them to make very wide swings from side to side.

ZOOPLANKTON, the animals which drift with the currents, also tend to sink as they are heavier (more dense) than water. They

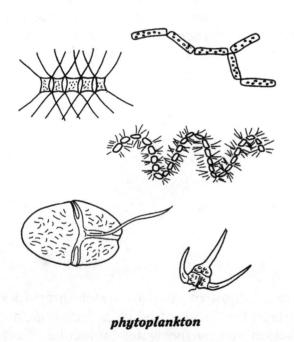

phytoplankton

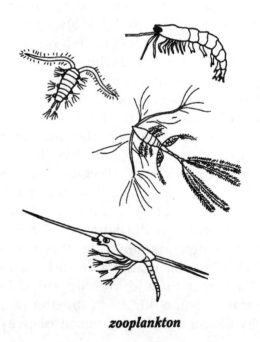

zooplankton

not only need to keep from sinking to the bottom, but also need to move to find food and avoid predation. Fortunately, they can often swim, though not strongly enough to swim against currents. While some remain near the surface, many zooplankton make a daily VERTICAL MIGRATION up to the surface at night and down into deep water during the day. They feed at night on phytoplankton or on the zooplankton that feed on phytoplankton. In addition to swimming, zooplankton also have structural adaptations that help them stay up. Many have long projections that create drag and slow the rate at which they sink. Zooplankton frequently have special organs which sense gravity so they know which way is "up." People typically think of zooplankton as tiny animals, but some are quite large, ranging from several inches to many feet. Some jellyfish are more than 100 ft long. Their universal characteristic is that they are drifters in currents even though they may swim enough to move up and down in the water.

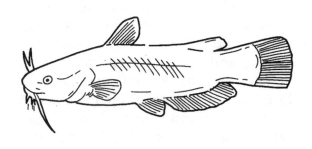

The flesh of all fish has a higher specific gravity (is more dense) than water. Without special adaptations, fish would sink, and indeed some live on the bottom. Bony fish that stay suspended in the water column have two ways to maintain their position. First, they may use their fins in swimming. A typical fish has seven fins.

The second thing that keeps many bony fish from resting on the bottom is an air or SWIM BLADDER. This organ is a pouch located between the stomach and the backbone. In primitive fish it is connected to the throat by a tube while in most fish it is sealed and gases enter and leave it from the fish's blood. It can be inflated or deflated by the fish to adjust the fish's BUOYANCY, keeping it "weightless" or neutrally buoyant at whatever depth it is swimming. The fish adjusts its overall density to that of water. The water it displaces equals its own weight—neutral buoyancy. If the water it displaced weighed more than the organism, it would rise and float—positive buoyancy. If the water it displaced weighed less than the organism, it would sink—negative buoyancy. This system works much like the buoyancy compensation device used by scuba divers to help them maintain their position in the water. The oil in phytoplankton serves the same function. The concept of buoyancy is difficult for middle school students. This entire curriculum can be taught without using the word.

Most bony fish possess a swim bladder, but the CARTILAGINOUS fish (sharks, skates and rays which have a skeleton made of cartilage) do not. Sharks swim continuously lest they sink to the bottom or live resting on the bottom. Sharks do have a huge, oily liver which helps make their overall density close to that of water since oil is lighter than water.

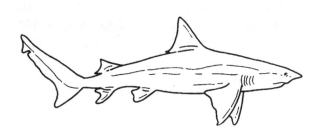

Viscosity and drag

Water also has VISCOSITY, which is resistance to flow. It is not as viscous as gelatin or pancake syrup, but it is much harder to move through water than through air. Animals that move through water have special shapes that help them slip through the water to reduce the energy they must spend on swimming. Plants and animals that are heavier than water sink more slowly than might be expected if they have structures such as projections or hairs that take advantage of viscosity by increasing the organism's drag or resistance to movement.

Dividing up the underwater world

These characteristics of water and the adaptations to them result in there being many different ways of life in different aquatic environments or HABITATS. A habitat is the place where an animal or plant normally lives. Aquatic habitats can be divided into three broad areas or places: the BENTHOS, or bottom for things heavier than water; the surface for things that float or rest on surface tension; and the WATER COLUMN, all the water in between occupied by organisms that establish near neutral buoyancy or swim to keep from sinking. Within these major areas are subdivisions categorized by the amount of light, vegetation, closeness to shore, kind of bottom, depth, food resources, and other plants and animals.

Animals and plants that live in aquatic habitats can be placed into COMMUNITIES that reflect where they live. Animals that live on the bottom sediment are in BENTHIC communities. Rooted water plants are also benthic and may be submerged (completely under water) or emergent. Those plants and animals that float or move on, in or just under the surface film of water are called NEUSTON and have features that help them exploit the surface tension that interfaces air.

PLANKTON are the small plants and animals that live in the water column, either drifting or weakly swimming. PHYTOPLANKTON (phyto = plant) are microscopic organisms that change inorganic nutrients into food and release oxygen when they do PHOTOSYNTHESIS. Like all plants, they can grow only in the EUPHOTIC ZONE (also called photic zone by some), where enough light penetrates for the process of photosynthesis to occur. Phytoplankton are eaten by ZOOPLANKTON (zoo- = animal) and by filter feeders ranging from clams and oysters to some small fishes both in the euphotic zone and as they drift below it toward the bottom.

The zooplankton community ranges in size from tiny, single-celled animals to larger creatures like jellyfish. It includes both temporary and permanent residents. Many freshwater and marine animals spend the early LARVAL stages of their lives as tiny drifting animals and later settle to become either slow-moving or SESSILE (anchored) adults or grow into free-swimming mobile animals, like fish. Strong swimmers that can move horizontally and vertically in the water column are called NEKTON. These animals can move against currents and tides to maintain their position.

ACTIVITY 29 To each its home

Where do animals and plants live, and what makes them suited to their homes? Classifying the places that things live in aquatic environments.

Teacher's information

This activity introduces the concept of different places to live within a body of water and the ways of life found in each. Saltwater and freshwater species are mixed as the basic subdivisions of space in each are the same. This exercise teaches the vocabulary needed for the following activities in this section. It is done by individual students who trade animal and plant Identity Cards to complete their Aquatic Homes Key. The vocabulary used in this activity is commonly found in oceanography books as well as books on freshwater habitats. The key progresses from places in the water to final answers that describe ways of life in those places. For example, the bottom is the place, but there are many ways of life for animals that live on the bottom.

Introduction

Draw a cross section of an ocean or lake on the blackboard or make a bulletin board as suggested. Ask enough questions about what you have drawn to make sure the students understand what the cross section represents. Ask the students to suggest some different kinds of animals and plants that might live there and where in the ocean or a lake they might live. Be sure to draw in some plants near the shore and some animals in the water and on the bottom. Do not introduce any terms yet. Let the students discover them in the key. Have the students reflect on the nature of a cross section as a means of communication.

Explain that animals and plants live in different places in aquatic environments, whether they are in the ocean you have drawn or a pond or river. Each kind of animal and plant has its own home and way of life in that home. Distribute the Aquatic Homes Key. Note that it starts with very broad divisions that get more precise with each level. Students go one step at a time, making a choice between two char-

Science skills
classifying

Concepts
Bodies of water are divided into different kinds of places with specific names.

Both freshwater and saltwater systems are divided into the same categories.

The characteristics of these places determine the kinds of plants and animals that live there.

Skills practiced
using a key
reading for information
building vocabulary

Time
1–2 class periods

Mode of instruction
independent individual or student pair work

Sample objectives
Students list the different places things live in an aquatic habitat.

Students classify living things by the location in which they live in an aquatic habitat.

Materials
for each student
Aquatic Homes Key
for the class
Identity Cards set
construction paper
optional
magazine pictures to illustrate the cards
butcher paper

Preparation

Make sets of Identity Cards by pasting copies of the enclosed cards on construction paper. Students may color the cards. You, students or parent volunteers may illustrate them by adding pictures cut from magazines that show the kinds of animals and plants identified on the card. Laminate the card set for durability. There are 16 Identity Cards, so you may duplicate sets. They mix saltwater and freshwater organisms on one card since the terms used are applied to both freshwater and marine habitats. You can list the kinds of animals and plants found in each location on the back of your Identity Cards. It is possible to do this activity without pictures, but not as much fun nor as instructive for students. If you enlarge the Ocean Cross Section, using an overhead projector to draw it on butcher paper for a bulletin board, the students can place their cards in the correct location in the ocean when they are identified.

acters each time, following the numbers to the next step. The Key assumes that students can distinguish plants (which make their own food) from animals (which move and eat other organisms).

Action

Pick one organism and use it as an example with the key. Have your students work with you to identify what kind of place it lives in and the name given to the group of animals or plants that uses that way of life. For example, plants that live completely under water and rooted in the mud (home) are called submerged aquatic vegetation (name for plants with that place and way of life).

Give each student an Identity Card. (Students may also work in groups at first.) Each is about plants or animals that live in a certain place or home. Students read the cards about the organisms that share this home and use the clues to figure out where they live. Fill in the card number in the space at the end of the key. Trade cards until each has inhabited all of the different places one can find within aquatic habitats. If you are short of time, do several cards and stop.

Results and reflection

Have your students name the different ways of life in the ocean used by the different groups of plants and animals they keyed out. Have them list the vocabulary words such as benthic or sessile they learned which name the different ways of life or aquatic homes. If you enlarged the ocean cross section, have the students place a card in the location where that group of organisms would be found. Again, discuss the nature and utility of a cross section for thinking and visualizing something that exists in three dimensions. Discuss its value as a representation of something very large.

Conclusions

Environments are divided into many ways of life that are used by different types of animals and plants. Where an organism lives is determined by many things such as the available food sources, light, depth and bottom characteristics as well as the mobility, weight, age and size of the animals and plants themselves.

Using your classroom aquarium

If you have several different kinds of animals and plants, the students can key them or make their own keys. Things to look for are: animals that stay near the bottom or clean the tank, such as catfish, *Plecostomus* and snails (benthic organisms); fish that swim in the water column, such as goldfish, guppies and mollies (nektonic organisms); plants that float at the surface (neuston); plants that grow rooted to the bottom (benthic). If you have both small and large fish, find their preferred spots in relation to real or fake plants.

Extension and application

1. Included with the set of Identity Cards is a short list of typical animals or plants that have those identities. You can glue these to the reverse side of the Identity Cards. You may also send your students on a picture hunt. After they have finished their key, have them choose one identity and check the list to find some organisms that have this way of life. Have your students research the plant or animal they choose from the list or pick another species that also has that way of life.

Outline

before class

1. make Identity Cards

2. copy Aquatic Homes Key

3. enlarge Ocean Cross Section

during class

1. introduce the idea of places in a three dimensional environment

2. demonstrate the use of the key with one card

3. let students work to identify other cards

4. place cards in correct locations on cross section

5. reflect on the drawing as a representation of reality

Examples of kinds of plants and animals in the Identity Cards

A. clams, tubeworms, some insect larvae, *Tubifex* worms

B. green algae, some seaweeds

C. most larvae (not fully developed forms) of fish, crustaceans, and echinoderms, and small tadpoles

D. horseshoe crabs, crayfish, flounder, anemones, *Hydra*, sea urchins

E. dinoflagellates, diatoms

F. mole crabs, sand dollars, quahogs and other clams

G. water striders, whirligig beetles

H. kelp, seaweeds

I. deep sea anglerfish, flashlightfish, and hatchetfish

J. copepods, cladocerans, rotifers, jellyfish, and krill

K. bladderwort, water hyacinth, duckweed, *Sargassum*, and *Hydrilla*

L. Portuguese man-of-war, "silver beetle," fisher spider, and backswimmer

M. barnacles, corals, sponges, mussels, and oysters

N. cattails, water lilies, pickerel weed, and saltmarsh cordgrass

O. most adult fish, squid, some sharks, and all marine mammals

P. turtle grass, water milfoil, seagrass, and pondweeds

IDENTITY CARDS

B. You look like scum or green hair waving in the water, and you feel very slippery. Wherever there is an available space on sunlit rocks or logs, or even floating leaves, you may be able to grow. You make food with sunlight.

D. You are not speedy. A hard shell, camouflage, stinging cells, or a place to hide may keep you safe as you move along the bottom. If you were lifted and then dropped, you might sink like a rock or might swim to hide on the bottom.

A. You are surrounded by dark mucky mud. Bacteria and fungi near you are working to decompose (rot) all of the dead animals and plants that sink to the bottom near you. You have tubes that reach out of the mud into the water to get oxygen rich water and bring in the food you filter out of the water.

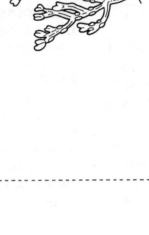

C. You live up in the water but are a very weak swimmer now, unable to swim against the current. You may grow out of this stage. Or you may spend your whole life in this community, drifting from one meal to the next.

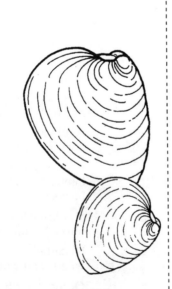

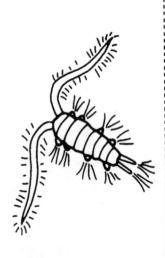

National Aquarium in Baltimore

IDENTITY CARDS

F. If you are on a beach, you may get uncovered by waves. If you are very small you can live between the sand grains, but if you are bigger, you may have a tube(s) to help you filter food and get oxygen from the water.

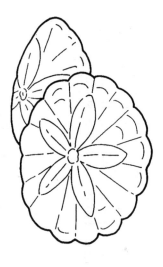

H. Many animals hide in your fronds or swim over them. You must live in shallow or clear water so that sunlight can reach you in order to make your food. You grow attached to rocks, pilings and other hard things.

E. When water looks very green, it is because of you. You may be tiny, but you are not alone. Others like you are drifting, slowly sinking. You are an important food source for small animals as well as the larger filter feeders.

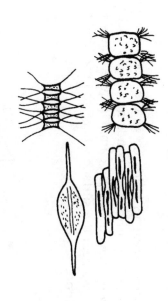

G. Special structures and body shapes help you to "walk on water." Wherever you step, you make small dimples on the water surface. You look down on things you might eat, but also have to watch out for animals above in the air that might eat you.

IDENTITY CARDS

J. Although you cannot swim against currents, you are able to rise or sink in the water, and do so. Your food is near the surface, but so are animals that might eat you, so you sink in the day to hide in dark waters and make your way up later to feed at night.

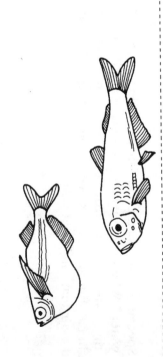

L. Your bubbles of air or oil are lighter than water and help you to stay at or just below the surface. You may even walk upside down on the surface. You eat other things that also live near the surface.

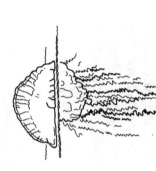

I. You swim in the deep. It is too dark to see, so you may use another sense to find food. Or if you spend your whole life here, you may even make your own "light" called bioluminescence.

K. Without roots to hold you to the bottom, you float wherever the currents or winds move you. Because of this, you often end up in a clump in quiet water. Floating keeps you near the surface where you use light to make your food.

IDENTITY CARDS

N. Waves or breezes may make your leaves move in the air but your roots in the water help to hold you in place.

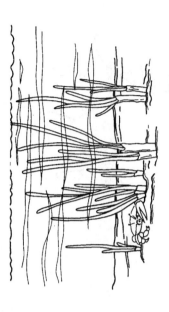

P. You live entirely underwater in shallow places where the light you need to make your food can reach. You are rooted in sandy or muddy bottoms. Baby fish and other animals may hide and feed among your leaves.

M. You might be found attached to anything from pier pilings to rocks and coral to boat bottoms. Wherever you are, you're there to stay and must wait for food to come to you for the rest of your life.

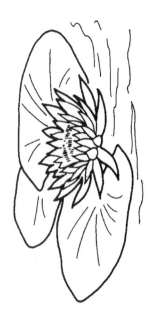

O. You can see the food you eat and can swim fast after it if it moves. You may live by yourself or travel in schools in open water.

National Aquarium in Baltimore

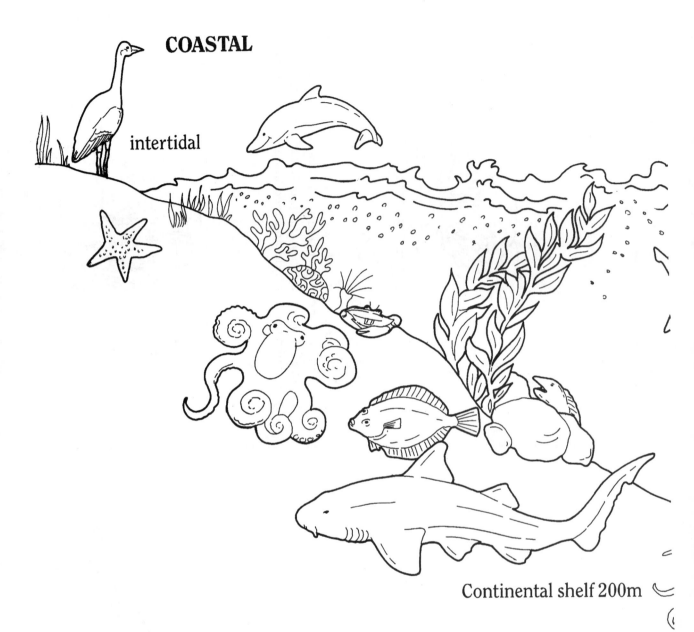

COASTAL

intertidal

Continental shelf 200m

benthic - bottom
nekton - swimming
plankton - drifting
   phytoplankton - plants
   zooplankton - animals

PELAGIC OCEANIC

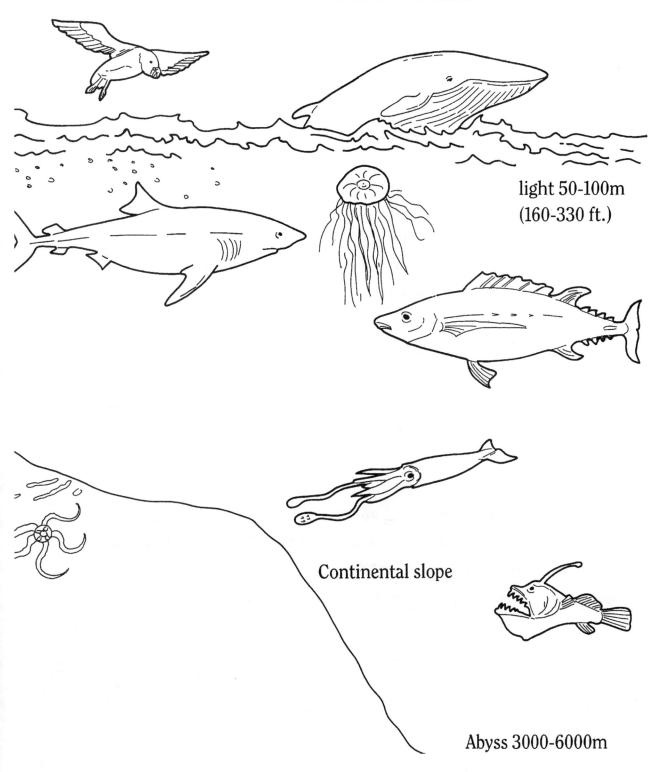

light 50-100m
(160-330 ft.)

Continental slope

Abyss 3000-6000m

Activity 29
Aquatic Homes Key

1. Plant 2

1. Animal 6

2. Lives on or in the bottom (BENTHIC) or lives attached to a surface 3

2. Lives up in the water, drifting or floating 5

3. Lives rooted in mud or sand bottom 4

3. Lives attached to a hard place: SEAWEEDS _____ and _____

4. Lives completely under water (submerged): SUBMERGED AQUATIC VEGETATION _____

4. Lives part out of water (emergent): EMERGENT PLANTS _____

5. Tiny single celled drifting plants: PHYTOPLANKTON _____

5. Large floating plants: FLOATING AQUATIC VEGETATION _____

6. Lives on or in the bottom (BENTHIC) 7

6. Lives up in the water above the bottom 9

7. Lives inside the mud or sand: (INFAUNA) _____ and _____

7. Lives on top of the bottom (EPIFAUNA) 8

8. Permanently attached to bottom (sessile): SESSILE INVERTEBRATES _____

8. Moves over the surface: BENTHIC ANIMALS _____

9. Lives up in the open water 10

9. Lives at or on the top of the water 11

10. Strong swimmer: NEKTON _____ and _____

10. Weak swimmer; drifts with currents: ZOOPLANKTON _____ and _____

11. Lives floating at the surface: NEUSTON _____

11. Lives up on the top of the water: EPINEUSTON _____

Activity 29
Aquatic Homes Key

Name __possible answers__

1. Plant 2

1. Animal 6

2. Lives on or in the bottom (benthic) or lives attached to a surface 3

2. Lives up in the water, drifting or floating 5

3. Lives rooted in mud or sand bottom 4

3. Lives attached to a hard place: SEAWEEDS __B__ and __H__

4. Lives completely under water (submerged): SUBMERGED AQUATIC VEGETATION __P__

4. Lives part out of water (emergent): EMERGENT PLANTS __N__

5. Tiny single celled drifting plants: PHYTOPLANKTON __E__

5. Large floating plants: FLOATING AQUATIC VEGETATION __K__

6. Lives on or in the bottom (benthic) 7

6. Lives up in the water above the bottom 9

7. Lives inside the mud or sand: (INFAUNA) __A__ and __F__

7. Lives on top of the bottom (epifauna) 8

8. Permanently attached to bottom (sessile): SESSILE INVERTEBRATES __M__

8. Moves over the surface: BENTHIC ANIMALS __D__

9. Lives up in the open water 10

9. Lives at or on the top of the water 11

10. Strong swimmer: NEKTON __I__ and __O__

10. Weak swimmer; drifts with currents: ZOOPLANKTON __C__ and __J__

11. Lives floating at the surface: NEUSTON __L__

11. Lives up on the top of the water: EPINEUSTON __G__

ACTIVITY 30 Keeping your head above water

Do things that float or sink behave differently in salt and fresh water? What lets them float? When do they sink? Student investigations.

Science skills
observing
measuring
predicting

Concepts
Because salt water weighs more (is more dense) than fresh water, it supports things more than fresh water.

Skills practiced
weighing
measuring liquid volume

Time
1 class period

Mode of instruction
teacher directed group work

Sample objectives
Students compare the way things float or sink in fresh and salt water.

Builds on
Activity 6

Teacher's information

BUOYANCY is a very difficult concept. This entire lesson should be taught without using the word. The idea that some things sink and others float is straightforward, but the reasons behind these observations are not so easy to accept. This activity allows students to experiment with floating and sinking, but does not go into details of displacement. An extension on displacement is included if you have very advanced students. It requires accurate measuring tools.

Introduction

Start with a discussion of your students' own perceptions of floating and sinking. Have they ever been swimming in salt water? Fresh water? Which was easier to float in? Have they ever been to the Great Salt Lake or seen people floating in it in pictures? How about the Dead Sea (also a salt lake)? In your class you should be able to find at least one student who has made the observation that it is easier to float in salt water than in fresh water.

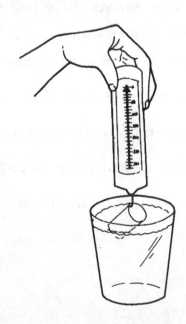

Action

Start with a challenge: can the students design an object that floats in salt water and sinks in fresh water? Let them experiment with film canisters and pennies (13 to 14 pennies in a plastic Kodak brand can generally works) in salt and fresh water in the measuring cups. To catch spills, place the cups in aluminum pans first. Does a film can holding the same number of pennies behave differently in two different solutions? How many pennies in a film can will float in fresh water? How many can salt water support?

What do they know about fresh and salt water from previous experiments? A volume of salt water weighs more than an equal volume of fresh water. Attach the 13 penny canister to the spring scale with a rubber band and lower it into each kind of water. What happens to the apparent weight? Can the students observe the water supporting the weight of the object? It should become "weightless" on the scale when it floats and lose weight even though it sinks. The water is supporting the canister.

Does a ball (about 1 inch diameter) of clay sink or float? It sinks. If the students change its shape, will it still sink? Try it flat or elongated in a bucket of water. It still sinks. Can the students figure out how to make it float? Forming it into a boat is the easy answer. Making it into a hollow ball is sneakier and much harder. Clay might be shaped around a ping-pong ball to make a hollow clay ball. It takes a great deal of trapped air to make the clay float. Has the weight changed on the boat or ball? Measure it. No. What has changed? Its volume. Its weight per unit volume has changed with the addition of air space.

Results and reflection

Salt water can float a heavier object of the same size (a more dense object) than fresh water. Have the students predict what would happen to a very heavily loaded boat as it sails up into a river from the ocean. It would sink lower and lower. Where harbors are shallow and have fresh water, boats have to be partially unloaded out to sea (a process called lightering) to keep them from getting stuck on the

Materials

for each group
four 35 mm plastic film canisters
50 pennies
250 gm Ohaus spring scale
2 clear plastic 2 cup measuring cups
   or large drink cups
rubber bands to hook to the scale
1 inch chunk of modeling clay sold
   in sticks like butter
2 aluminum pans

for class
fresh water in plastic jugs at room
   temperature (1.5 cups per group)
very salty water (1.5 cups table salt
   per quart) in plastic jugs at room
   temperature (about 1.5 cups per
   group)
bucket of fresh water

optional
ping-pong ball

Preparation

Mix the salt water the day before, using hot water to dissolve the salt. Let it sit to reach room temperature.

Outline

before class
1. mix salt solution

during class
1. have the students discuss what they know about floating and sinking

2. have the students use film cans and pennies to design something that sinks in freshwater and floats in salt water; discuss

3. have the students lower a can with 13 pennies attached to the scale into fresh and salt water; what happens to the apparent weight?

4. experiment with balls of clay; do they sink or float?

5. if it is flattened, does the clay sink or float?

6. can the clay be shaped into a hollow ball (around a ping-pong ball?) that floats? Has the weight changed? What has?

7. discuss observations

bottom as they sail up from the ocean. Have the students noticed marks painted on big ships that tell how low they are sitting in the water?

Conclusions

Salt water supports more weight than fresh water. Things float more readily in salt water than fresh water.

Extension and application

1. If you have older and/or gifted students, you may also introduce the concept of DENSITY by having the students calculate the weight per unit volume of objects. You also need accurate graduated cylinders and good balances, not the spring scales. Weigh each object on a balance. Find the volume of the objects by filling a larger graduated cylinder part way with water with a bit of detergent to break the surface tension. Record the level. Then sink the object below the surface and record the new volume. Subtract the volume of the water from the volume of the water plus object to find the volume of the object. When all the objects' densities have been calculated, arrange them in order. What is the density of the fresh water? The salt water? Weigh a measured volume to find out. Where do they fit in the series of densities of the objects? Can you make a statement about density of an object versus density of a fluid with regard to whether it sinks or floats? If the object is less dense than the fluid, it will float. If it is more dense, it will sink.

2. If you have great students who like to think about space travel and planetary science, you may have your students calculate SPECIFIC GRAVITY for each object. Weight depends on gravity. Things weigh less on the moon where the gravity is less than on earth, but they have the same MASS. Relative mass can be expressed as specific gravity. Specific gravity generally uses distilled water at 4 °C as a standard and sets it equal to 1. Everything is compared to it. You could use cold tap water without being too

far off. Divide the density of an object by the density of the fresh water to get the object's specific gravity. For example, if the object were 2 gm/cubic centimeter (milliliter) and water is 1 gm/cubic centimeter (milliliter), then the specific gravity of the object is 2. The units cancel out. Anywhere in the universe the specific gravity of that object will always be 2 although the object's weight will change with gravity.

3. Specific gravity of liquids is measured with an instrument called a hydrometer. Students can make a hydrometer, using a thumb tack in the end of a pencil or clay holding BBs in a soda straw. You may also have a real hydrometer. Once they have made one, students can compare the specific gravity of other liquids to water by comparing how high the hydrometer floats in different solutions. They can also compare salt water or hot water. If liquids are used that are dangerous (i.e., methyl alcohol), be sure that students are aware of the possible danger or, better yet, do not use them. Also, note that methyl alcohol "melts" some plastic graduated cylinders. Household products like cooking oil, pancake syrup and corn syrup should do.

4. Invite interested students to investigate the Dead Sea. Students should ask such questions as:

    • Why is it called the Dead Sea?
    • Is it really "dead"? If not, what lives in it?
    • How did it get the way it is with very high salinity?
    • How do people who live around it use it?

ACTIVITY 31 Sinking slowly

How do tiny plants, which need light, maintain their position in the water column if they are heavier than water? How do tiny drifting animals keep from sinking to the bottom? A contest to design an organism that sinks the slowest.

Science skills

observing
measuring
inferring
predicting
experimenting
communicating

Concepts

Phytoplankton and zooplankton have a variety of adaptations which help them remain in position in the water column.

Skills practiced

measuring time
building models

Time

1 class period plus homework or 2 class periods

Mode of instruction

independent individual work

Sample objectives

Students describe strategies plankton use to maintain their position in the water.

Students apply an understanding of form and function to the construction of a model.

Builds on

Activity 29

Teacher's information

Students will observe pictures or a film about plankton and then apply what they have learned to the construction of a model. The goal in constructing a model of phytoplankton or zooplankton is to make the one that sinks most slowly. The construction phase may be done as a homework assignment with the contest taking place at school, preferably outside.

An extremely common misconception among both children and adults is that phytoplankton and zooplankton float. This error also appears frequently in curriculum materials, books and magazine articles. As students learned in Activity 30, things that float, by definition, are at the surface of the water. They are less dense than water. If phytoplankton or zooplankton floated, they would exist only at the surface of the water. There is a name for the few organisms that do float, such as the Portuguese man-of-war: NEUSTON. Phytoplankton and zooplankton are very slightly more dense than water and tend to sink, which is the problem this exercise addresses. Both groups have adaptations that enable them to remain up in the water column. Both are close to neutral buoyancy. Both frequently have projections that create drag (like a parachute does in air), slowing their descent. Zooplankton are often capable of swimming sufficiently to move up and down in the water, sometimes over great distances on a daily cycle.

Introduction

Start with observations of zooplankton and phytoplankton. Observe their shapes, projections and behaviors. Plankton are often the exact opposite of streamlined. Plankton are slightly heavier than water and tend to sink. Ask the students to speculate on how they might stay up in the water. Make a list of the students' observations. Encourage thought about the discoveries on density from previous activities.

Some of the students should notice that many plankton have long projections or antennae or hairs. Have them speculate on how these would affect movement through water. Could the students run through water faster with their own arms spread out or folded up? What tactics would they use to jump into the water if they did not want to have their head go under: the giant stride which presents the maximum surface area of spread arms and legs as one jumps in. Conversely, if they wished to go under water when jumping in, they would present the most streamlined body shape to get as little resistance as possible. Think of a good dive with the streamlined entry.

Action

Now the students are going to see if they can make a model phytoplankton or zooplankton which will sink slowly. Since "thrashing" or swimming is not possible in a non-mechanical model, they must concentrate on designing a plant or animal that is just barely heavier than water. Their design should slow its rate of sinking by increasing its resistance to movement through water with long projections or hairs. Alternately, it might sink slowly because it swings back and forth as it goes down. Have selections of materials and

Materials

for each student
worksheet
for each large group
stopwatches or digital watches which read in seconds
bucket of water
for class
pictures of phytoplankton and zooplankton
movie or video on plankton if available
Plankton and the Open Sea, 18 minutes, Encyclopedia Britannica
Plankton of the Sea, 12 minutes, Fleetwood Films
Plankton: Pastures of the Ocean, 10 minutes, Encyclopedia Britannica
Plankton: the Endless Harvest, 18 minutes, Universal Education
live or preserved plankton if available (see Recipes for ways to view plankton and for sources)
junk for construction: clay, plastic vials, nuts, nails, toothpicks, wire, strings, Styrofoam pieces, cooking oil, film cans, aluminum foil, coffee stirrers, straws, glue
a large trash can full of water
small, inexpensive prizes

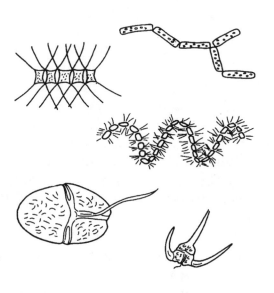

phytoplankton

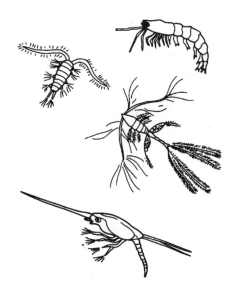

zooplankton

Preparation

This is a great, messy activity. It is particularly good for a warm day when the class may be naturally restless and ready for some excitement. The finish is a "slow sinking" contest so collect some prizes or even "coupons" for lost homework. Let students work overnight or over a weekend to perfect their plankton model. It can be tested in a bathtub or the kitchen sink.

Outline

before class

1. collect the junk

2. fill the trash can

3. copy worksheet

during class

1. introduce plankton through film, pictures, live ones, or drawings

2. discuss the adaptations phytoplankton have which help them stay up in the water

3. let the students experiment with designs for slow sinking plankton

4. have class establish rules for the designs

5. test and refine designs in buckets

6. hold a sink-off

7. stop at the last two and have the students predict final winner and give reasons based on the design; then test

8. compare designs and relative success; which designs work and why; no worksheet with answers is given due to great variation in designs.

buckets of water available around the room for design and testing. Use stopwatches to time the speed of sinking. Set a time limit for experimentation. After experimenting for a bit, have the class write a list of rules for the designs (i.e., no fabric or paper, size limit) and a list of rules for how they items will be released. Leave the rules on the board. Hold a contest for the slowest sinking animal or plant at the end of that time. Alternately, have the students design their plankton at home and test them in the bathtub. Outlaw paper or paper products of any kind. A dollar bill wins every time otherwise!

Now have the contest: a sink-off. Have each group determine the slowest within the group. Then have the group winners sink against each other in a big trash can of water or a very large bucket. Beware of over stressing a cheap plastic trash can which may pop. If you cannot give everyone a good view for all, have several students help you as judges. You can time each separately, but it is more exciting if pairs of phytoplankton or zooplankton are released by you to "reverse race" their way down. The sink-off builds excitement for the final "slow-down." The technique for releasing them is important. Hold them just under the surface of the water. Any model that pops up is disqualified because it floated. Release them at the same time for a fair start. The **slowest** from each pair goes into a second heat and so on until you get down to two. Stop at two.

Results and reflection

Have the students analyze what they think made each of the last two models winners. Then have them vote on which they think will win the grand prize for slowest overall based on their analysis. Do the final test and distribute prizes. One word of caution—beware of models that sink because they are gradually filling with water or absorbing water as paper or fabric does. That is why you hold the objects under water to release them. Anything that rises before sinking floated and then took on water. They must be heavier than water to start with. Something just barely heavier than water with lots of projections should win unless a student can produce a flat, pie pan shaped object that makes big swings from side to side as it descends.

Have the students reflect on the thinking they used in inventing their model plankton. Depending on the specific models created, you may wish to ask such questions as:

- Why did you use a plastic vial rather than a film can?
- Did you visualize your model before you started it? How did your mental picture of the model change as you worked?
- What effect might there be on your model if you used a toothpick rather than a nail right there?
- Why didn't you make the whole thing out of thin wire?
- Where did you get the idea for your model?
- How is your model like real plankton?
- How is your model different from real plankton?
- How did you decide when your model was finished?

No sample answers were given as the range of possible appropriate responses is too wide.

Conclusions

The strange looking shapes of phytoplankton have specific functions, including helping them stay in the lighted zone of the water where they can get enough light for photosynthesis (making food using light energy). Zooplankton shapes help them stay where they can catch food.

National Aquarium in Baltimore

Name _____

1. Draw a picture of the model plankton you designed.

2. If you did not have the slowest sinking model, compare how your model differed from the winner in terms of its shape. Describe how you would change your design to make it sink more slowly in the future.

ACTIVITY 32 A tense place to live?

What is the surface of the water like? Can animals take advantage of the structure of the water surface as a place to live? Student investigations of surface tension.

Teacher's information

These exercises examine the tendency of water to stick to itself (see Section IV Introduction). This cohesive property of water is perhaps most obvious at the surface. The top layer of water molecules forms a film or "skin" which is relatively strong caused by SURFACE TENSION. Many animals and plants live directly on the surface of bodies of water. Even though they are heavier than water and cannot float, they stay at the surface by "riding" on the surface film. In these activities students will experiment with surface tension. In Activity 33, they will design an animal that is heavier than water, but stays at the surface because of the surface tension of water.

Introduction

Refer to earlier activities in which students used plastic spoons to layer water in a container. Ask students if they noticed how the water in the spoon formed a nice, rounded top. Have them fill a plastic spoon with water and observe it from the side. What causes this? Let them find out.

Action

Have each group put a clear plastic glass in a small pie pan and fill the glass until the water is exactly level with the top of the glass. Would they say the glass was full? Yes. How many florist's glass balls (used to hold cut flower stems in a vase) or marbles or pennies do they think they could put in before the glass runs over? Have them estimate and record the number. Now slip pennies or marbles into the cup one at a time, counting and recording as they go. What does it appear happened? The water piles up into a rounded bulge until the surface finally appears to break. How many glass balls, marbles or pennies did it take? Have each group write its numbers on the board and then plot the results. Then

Science skills

observing
predicting
experimenting

Concepts

Water has a strong, elastic surface which forms due to surface tension, caused by the cohesion of water molecules.
Some animals live on the surface.

Skills practiced

hand/eye coordination
graphing
averaging

Time

1 class period

Mode of instruction

teacher directed individual and group work

Sample objectives

Students demonstrate the existence of surface tension.

Builds on

Activity 29

Materials

for each student
clean plastic spoon
clean plastic cup (clean, no soap residue)
2 small paper clips

for each group
small pie tin
water
50 pennies or 20 florist's glass balls or marbles
eyedropper or cotton swab
1/2–1 gal plastic deli container or bowl
1 tbsp dishwashing detergent in a small cup of water

Preparation

Make sure all the cups, spoons and deli containers have been well rinsed with non-soapy water and make the dilute detergent solution—about 1 tbs/cup of water.

add the numbers and calculate the average number of marbles or pennies dropped in before the rounded top was "stretched" too far.

Now discuss what happened. Water molecules are attracted to each other: they stick together. At the surface these molecules produce a layer that acts like a "film" that covers the surface and holds it together due to this cohesion which we call SURFACE TENSION. When the weight of the water pushing against water molecules was great enough, they broke apart and the water fell over the sides of the glass.

How strong is this surface cohesion of water molecules? Would the students predict that a paper clip would float or sink? Try it in a clean cup of water. Let them experiment to see whether they can rest a paper clip on the surface of the water. Help only if they are all failing and becoming frustrated. The illustration shows how a paper clip can be lowered onto the surface so it rides on the surface. After they have tried, if none of the students are successful, demonstrate this method and let them try it.

Can you prove that the paper clip is not floating, that it is not lighter than water? Touch it gently and it

sinks. Rest a paper clip on the surface and add a drop of dilute detergent to the water. The paper clip sinks even though it was not touched. Detergent destroys the surface tension of water.

Results and reflection

The students generally make very low predictions for the number of objects that can be added to a full container. Why did different groups get different results? They may have started with slightly different amounts of water; the way they dropped the objects may have been different; or the glasses may have been a bit dirty which changed the surface tension. It is normal for there to be variability in results which is why experiments must be repeated over and over to be sure. The use of statistics enables us to compare variable results.

Ask the students to write a paragraph explaining in their own words why they could place the paper clip on the surface of the water and it did not sink. They may also draw diagrams. Examine the student responses carefully to make sure they understand that the surface layer that supports the paper clip is composed only of water molecules.

What kinds of animals might be found living on the surface of water? We learned in Activity 29 to call them EPINEUSTON as a category. Examples listed included water striders and whirligig beetles.

Conclusions

The surface layer of water molecules has strength. It can be seen to stretch before it is broken. This layer can support things that are more dense than water on the water's surface. The objects are not floating, because they sink when detergent is added to break the film.

Extension and application

1. Begin a discussion of the creatures that live on the surface tension and their special adaptations for this environment. Have students research the animals such as water striders—animals that

Outline
before class
1. rinse cups thoroughly
2. copy work sheet

during class
1. let students observe rounded water on a spoon

2. fill a cup full to level with top with water

3. predict how many pennies or glass florist's balls can be added before it spills over

4. try it and record numbers

5. discuss observations

6. discuss source of variations in observations

"walk on water." They might be surprised to learn about spiders that live on the surface of the ocean. Most amazing are young *Basiliscus* lizards which can escape from predators by running across a river on the surface layer of water molecules, using the long toes of their hind legs. These lizards are common in central American rain forests. Use pictures the students draw along with descriptive paragraphs for a bulletin board.

2. Discuss strategies for reducing the local mosquito population. Since the larvae develop while hanging upside down on the surface tension and use a snorkel-like breathing tube, spreading cooking oil on the water's surface to block their air supply can act as a "natural" pesticide. What harmful effects to other plants and animals may be the result of this practice? Oil can prevent oxygen from diffusing into the water.

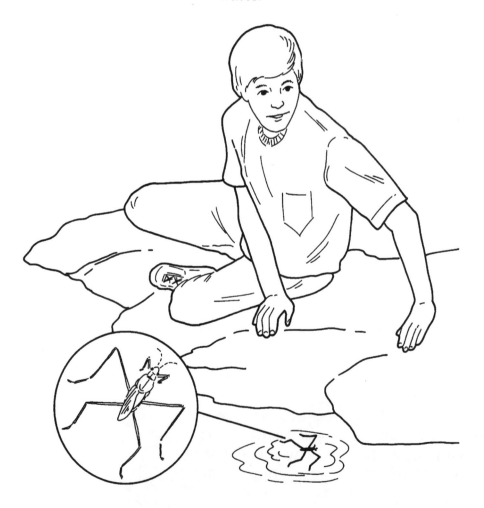

Activity 32
A tense place to live?

Name _____

A. Adding objects to the full cup of water

1. What kind of objects are you adding to the full cup of water?

2. How many objects do you predict that you will be able to add to the glass before the water runs over? _____ How many did you actually add? _____

3. Describe what you observed about the water when you added objects to a full glass of water?

4. On the board, working with other groups, make a bar graph that shows the number of objects each group added before their water ran over.

5. What is the average of all these numbers? _____ What is the range? _____

6. List two things that might account for the groups getting different results when they did the same experiment.

B. Resting the paper clip on the surface of the water

1. Write several sentences and use a diagram to explain how the paper clip was supported by the surface of the water.

2. Describe what happened when the detergent was added to the cup with the paper clip.

Activity 32
A tense place to live?

Name __possible answers__

A. Adding objects to the full cup of water

1. What kind of objects are you adding to the full cup of water?

   We are adding pennies.

2. How many objects do you predict that you will be able to add to the glass before the water runs over? __6__ How many did you actually add? __47__

3. Describe what you observed about the water when you added objects to a full glass of water?

   We just kept on adding pennies. The water bulged up, but it didn't spill down the side. Finally, it looked like the surface of the water "broke" in one place and it ran over the side of the cup.

4. On the board, working with other groups, make a bar graph that shows the number of objects each group added before their water ran over.

5. What is the average of all these numbers? __37__ What is the range? __23–56__

6. List two things that might account for the groups getting different results when they did the same experiment.

   1. We may have had a different idea of what full meant when each group filled its cup. We may have started with different amounts of water.

   2. We may have used different ways to drop or slide the pennies into the water. The group with the lowest number dropped the pennies in from way above the water.

B. Resting the paper clip on the surface of the water

1. Write several sentences and use a diagram to explain how the paper clip was supported by the surface of the water.

   The paper clip is heavier than water. We slid it carefully out onto the surface of the water so it did not mess up the surface tension of the water. Because we did not disturb the surface, it rested on it.

2. Describe what happened when the detergent was added to the cup with the paper clip.

   The surface appeared disturbed and the paper clip sank immediately.

ACTIVITY 33 Life at the surface

A student contest to design a surface-dwelling model animal.

Teacher's information

Be careful not to refer to things that are supported by surface tension as floating. They are heavier than water and do not float. Try using the concept of riding or resting on surface tension.

Introduction

Following Activity 32, ask the students about living things that use surface tension. Have any of the students ever seen animals that use surface tension? Some kinds of beetles and bugs walk on water in search of prey. The water strider, a familiar pond bug, has special hairs on its first and third pairs of legs which form dimples on the water surface. The strider's second pair of legs actually penetrates the surface and works like oars to propel the insect over the surface. There is a marine species of water strider as well. Other kinds of insects, like mosquito larvae, hang upside down from the surface film and poke a breathing tube up through it. A pond insect called a springtail has a spring-like appendage with which it jumps around on the surface of ponds and temporary water holes. Whirligig beetles are so well-adapted for life on the surface tension of ponds that each eye has two halves: the upper half can see above the surface while the lower half simultaneously views the underwater world! Some insects that might otherwise be preyed upon by water striders are able to spit small amounts of detergent into the water and cause the strider to go under before grabbing its meal!

Action

Can your students design an animal that uses surface tension, living on the surface of water even though it is heavier than water? This is very difficult and may be frustrating. The winner is the student who designs the heaviest model organism that rests on the surface but is heavier than water. Provide them with very thin wire and other materials that are heavier than water with which to build their "creatures." Remind them that the models must sink when detergent is

Science skills

observing
predicting
experimenting

Concepts

Models allow scientists to test their understanding of how things work.

Modification based on observations and predictions enables perfection of models.

It is possible to design things heavier than water that rest on the surface of the water.

Skills practiced

building a model

Time

1 class period

Mode of instruction

individuals or pairs; independent classwork or homework

Sample objectives

Students create a model of a living organism from simple materials.

Builds on

Activity 29
Activity 32

Materials

for each student
worksheet
for the class
very thin florist's wire (hobby shop or dime store)
other materials heavier than water such as rubber bands or pins
1/2–1 gal deli containers
optional
pictures of water striders

Preparation

Try making the model yourself. Very thin wire makes a nice "water strider." This is not easy. Making looped feet like small snowshoes helps. The students will learn more if they are allowed to work through trial and error rather than by copying your design.

Outline

before class

1. copy worksheet

2. collect materials

3. try a design yourself

during class

1. review Activity 32

2. discuss animals that live on the surface of water, using surface tension

3. challenge the students to design their own surface-living animal that rests on the surface, but is heavier than water and does not float

4. allow students to experiment with designs

5. test models

6. reflect on designs and on the thinking that was behind the designs; how did they think about doing the project?

added to the water, yet must rest on the surface when placed gently. Let them select from among your materials or add from their own collections. Give them a set period of time to complete the project, such as over a weekend.

Results and reflection

You might want to turn this exercise into an elaborate contest. If so, have the students with viable entries bring their creatures to the front of the room. Allow them to place their models, one at a time, on the surface of a large, clear container of water in full view of the class. Weigh the ones that are successful and determine the heaviest. Rest all of them on the water after weighing and add detergent. Only those that sink are eligible to win. The rest were floating. Award a prize to the winner.

Initiate a follow-up discussion concerning which model shapes and parts worked best. It will probably be clear that the models with their weight evenly distributed over the surface area, not all in one spot, work best. Have your students reflect on the way they thought about doing this project. What mental approach did they use for inventing their creatures? No sample answers are given due to diversity of possible responses.

Conclusions

While water surface tension is exploited by some animals, only a few designs work. It is very hard to rest on the surface. Only designs that spread the animal's weight over a very wide area work. This can be likened to what happens when people try to walk on the thin ice of a pond. They break through. To save themselves they may crawl to shore by spreading their arms and legs while lying flat and inching along carefully. They do not weigh less, but have spread their weight over more of the ice surface.

Using your classroom aquarium

If any of your students live near a slow creek or pond, ask if they can safely collect some water striders. Add them to your aquarium while all the students are watching. In addition to seeing animals ride on surface tension, they may get to see how the water striders deal with hungry fish.

Activity 33
Life at the surface

Name _____

1. Draw a picture of your model of an animal that lives on the surface of the water.

2. Did it really rest on the surface? If it did not stay up or if it floated, describe how you might change it to make it work better.

3. If you did not win, compare your design to the winning one. List two ways the winner had a better design than yours.

ACTIVITY 34 Sink or swim

What are the special structures that allow fish to stay up in and move through the water? A fish dissection.

Science skills
observing

Concepts
Most fish are well-adapted as swimmers.

Fish have special structures such as fins which enable them to move through the water.

Most fish have a swim bladder which helps them maintain their position in the water by adjusting the fish's density to near that of the surrounding water.

Skills practiced
dissection
following written instructions

Sample objectives
Students do an orderly dissection to investigate a question.

Students list and locate the fins of a typical fish.

Students locate the internal organs of a fish.

Time
1–2 class periods

Mode of instruction
independent group work

Builds on
Activity 29
Activity 30

Teacher's information

Observing the external and internal anatomy of a BONY FISH is one way to answer the main questions that this lesson poses: how do fish move through the water and how do fish keep their vertical position within the water? Your students will also learn about other fish adaptations and characteristics. Contrary to what students may think, a dissection is a clean, neat and organized activity. It is not a cut, slice, hack and saw proposition; rather, each part must be carefully unwrapped from the other organs nearby. It should be thought of as uncovering the layers, one by one, so as not to harm the layer below. It is important to stress this with the students before beginning any dissection. Simple scissors with one sharp point and one rounded point are the only cutting tool needed. **Note:** no answer sheet is given because fish are too diverse to write one that is useful.

Introduction

Review special adaptations some organisms have for living in different places in the water. Some have special appendages to slow their rate of descent through the water; others have structures that allow them to live on the surface tension. But what about fish? What do they have that keeps them from sinking and enables them to occupy different levels in the water? Students will no doubt mention fins, but is there anything else? How about studying the external and internal anatomy of a fish to find out?

Action

Discuss the safe use of scissors and the rules for behavior during dissection. **Warn students to handle the fish carefully, looking for sharp spines in the fins.** Students tend to want to stick their fish in somebody's face to hear screams. Pass out the fish and worksheets and let teams proceed independently while you circulate to keep things going.

Results and reflection

Review student observations of external anatomy. Compare fish if you had more than one species. Did they all have the same number of fins? Were they located in different places? Did the fish all have the same shape? Have the students speculate on which shapes might move through water most easily. When the students have completed their dissections, ask if they found a swim bladder as well as other structures. If you had bottom-dwelling species like a catfish or a flounder, the students did not find one because they lack one. Fish that rest on the bottom are heavier than water and swim actively in order to get off the bottom.

How does a swim bladder work? Try this demonstration. Ask students to predict what will happen when a dead fish, that has already been frozen and thawed, is placed in a tub of water. Try it! If it is not rotten, it will sink. Now, make a small cut with scissors in the underside of the fish (beginning at the vent opening). Insert a very small balloon, inflated just a little, into the opening. Use a needle and thread or safety pins to close the cut and hold the balloon in place. Does the fish sink or float when placed in a container of water now? This demonstration shows how a fish's swim bladder actually works to keep the animal suspended in the water, although its actual position will not be correct. You might also try taping a balloon to the fish's back for a demonstration of balance. No answer sheet is given due to the diversity of possible responses based on kind of fish used.

Conclusions

Fish use a combination of fins and a swim bladder, if they have one, to maintain their position within the water. Fish that rest on the bottom are heavier (more dense) than water while fish that live up in the water can adjust their density to that of the water around them.

Using your classroom aquarium

Observe live fish swimming in your aquarium. Note which fins are used, how they are used, and how easy or difficult it appears for different species of fish to remain suspended in the water. Do some

Materials
for each group
small fish (10–14 inches); not cleaned
newspapers or a dissecting tray
one pair small sharp scissors (one point side pointed, one rounded)
one probe (coffee stirrer, round wooden stick or sturdy toothpick)
one pair of tweezers (useful but not required)
paper towels
worksheet
drawing of fish's insides
for the class
dissecting microscope, hand lens or magnifying glass
one fish about 1 ft long for demonstration
small balloons
an aquarium tank or bucket of water
safety pins or needle and thread

Preparation
The best fish for dissection are ones you catch and immediately freeze. Put out a call among your students' parents to freeze small fish (10–14 inches) from a fishing trip. If treated properly, they will be fresher than grocery store fish. You might get more than one kind of fish. One author has used as many as 10 different kinds of fish in a class. Different kinds enable the students to make interesting comparisons of anatomy and to speculate about the function of these differences. Place the fish on ice immediately upon being caught and in the freezer as soon as possible. The insides turn to mush if the fish are allowed to sit around. Fresh fish from the grocery store are not as fresh as one would wish but will do. If frozen, thaw in cold running water shortly before class. Saltwater fish frequently have a larger swim bladder than freshwater fish. Practice dissection and prepare a fish to serve as a model for the students by dissecting it and labeling the internal organs in advance. If you plan to do the fish printing activity, do it **before** the dissection.

Outline

before class

1. locate fish

2. thaw under cold water if needed

3. practice dissection

4. label parts on dissected fish for student reference

5. if you are doing fish printing, do it first

during class

1. review other organisms' adaptations

2. discuss process of safe, orderly dissection

3. let students follow the worksheet

4. help as needed

5. review and discuss observations

6. demonstrate the swim bladder's function with balloon

7. discuss in relation to floating and sinking

remain motionless, hanging in the water? Which fins are used for swimming? Do all your fish swim in the same manner?

Extension and application

1. During dissection, have students remove sections of the backbone and/or the jaws and teeth. These can be cleaned, dried and preserved for later observation by removing as much of the flesh as possible, placing the bones in a jar of bleach for several days and then setting them somewhere safe to air dry.

2. Show the film *What is a Fish?* This 22 minute Encyclopedia Britannica film #2033 includes footage of a dissection as well as sections on adaptations. Even though this is an ancient film narrated by a stereotypic male nerd scientist, it is excellent.

3. Compare the differences, both internal and external, between sharks and bony fish by dissecting a shark with your students. Do only one as sharks are slow to reproduce and should not be used heavily. Sharks can be obtained from biological supply companies, fishermen, and sometimes fish markets if you let them know you're looking for one. Several manuals exist concerning shark anatomy and are available from biological supply houses. One major comparison would be the large oily liver of a shark, which serves a function similar to the bony fish's swim bladder in making the shark less dense. Place a piece of shark liver in a bowl to demonstrate that it floats. Some sharks also swallow air at the surface for temporary buoyancy.

4. A good addition to a fish anatomy lesson is an art activity called fish printing. See the Fish Printing page for instructions. Work with the art teacher in collecting the materials for this activity. This activity also fits into social studies if you are studying Asia.

5. No sample answers are given because of the diversity of possible answers with different fish.

Fish Printing

The Japanese invented "gyotaku" (gyo = fish, taku = rubbing) as a means of recording their catch. (A fish print never lies!) Gyotaku (pronounced ghe-o-ta-koo) has since evolved into an art form and is a good way to gain appreciation for the beauty and variety of marine organisms. Fish printing may be done on paper or fabric; directions are similar and given for both.

Materials

> fish (any fish that is fairly flat and has obvious scales)
>
> newspaper
>
> paper towels
>
> newsprint (rice paper is traditional but expensive)
>
> water soluble paint or textile ink
>
> stiff-bristled paint brushes
>
> thumbtacks or tape

This can be rather messy and time-consuming, so plan accordingly. If you use textile paint, the students may print a T shirt or other item. Use cotton cloth for best results. Do a quick demonstration for the class to illustrate the preparation, speed and amount of paint needed. Here are the steps:

1. Before the class, wash the fish carefully and thoroughly with soap and water to remove the mucus. Plug the mouth and gills, and vent (if possible) with paper towels so the fish will not leak.

2. Paint the fish. Do not to use too much ink. Thin the ink or paint to the consistency of cream if necessary. Stroke the brush from head to tail, but do not paint the eye. Paint fins and tail last, since they tend to dry out quickly. You may also brush the fish from tail to head to catch ink under the edges of scales and spines and improve the print if you use a thin coat of paint.

3. If the newsprint paper under your fish became wet with ink during the painting process, move the fish to a clean piece before printing. Otherwise your print will pick up leftover splotches of color.

4. Gently lay a sheet of paper over the fish. Taking care not to move the paper, use your hands to press the paper over the fish. Press the paper gently over the fins and tail. Be careful not to wrinkle the paper or you will get a blurred or double image. If printing a shirt, put a layer of newspaper inside the shirt before printing so that the paint does not seep through to the other side.

5. Slowly and carefully peel the paper off. Paint in the pupil of the eye with a small brush. The prints tend to all look alike, so have students write their names on them and tack or tape them to dry.

6. Once a fish has been painted, it should be washed and dried before changing colors. If this is not feasible, use paper towels or a damp sponge to remove as much paint as possible.

7. Fabric items must be ironed (cotton setting) from the reverse side for 30 seconds to set the color first and then washed in mild detergent.

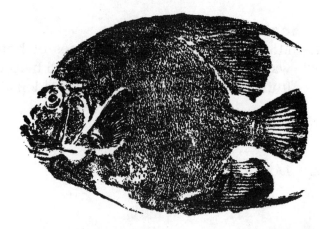

Activity 34
Sink or swim

Words that are vocabulary words associated with fish anatomy are written in CAPS.

I. The outside
A. Shape
1. Describe your fish's general shape.

2. Draw your fish in the space below and **label all its fins and other parts** as you do this worksheet.

B. Skin
1. Feel the surface of the SKIN. Describe what it feels like.

2. The skin is covered with MUCUS which helps protect it from disease.

C. Scales
1. Most bony fish have tiny pieces of bone in their skin called SCALES. Scales help waterproof the fish's skin and also help protect the fish. Remove a scale and observe it under the microscope or a hand lens. Draw the fish's scale here.

2. Does the scale have rings?

3. If so, how many rings can you count?

4. What might the rings tell you?

5. Can you name any other living things which have rings?

D. Lateral line

1. Can you locate a line down the side of your fish? Most fish have a LATERAL LINE which senses movement in the water near the fish. Label the lateral line on your diagram if your fish has one that can be seen.

E. Nares

1. Can you find two little holes in front of the eyes? These are called NARES and provide the fish with a sense of smell. Mark them on your drawing. Unlike your nostrils, there is no connection to the fish's mouth.

F. Fins

Find and label these FINS in your drawing.

1. The tail fin is called the CAUDAL FIN.

   a. What do you think might be this fin's main use?

   b. A narrow, forked tail means your fish is a constant, fast swimmer. A broad tail gives good power, but in short bursts. Look at your fish and decide which kind of tail your fish has.

2. DORSAL FIN(S) are those on the fish's back. In most fish these fins help keep the fish upright in the water, but some fish move their dorsal fins in an s-shaped wave to swim. Fish can raise or lower their dorsal fins which may lie flat on fast swimmers.

   a. How many dorsal fins does your fish have?

   b. Does your fish have SPINES (hard and sharp) or RAYS (soft and flexible) in its dorsal fins or both? Spines can protect the fish by sticking a predator's mouth.

3. PECTORAL FINS are the pair of fins on either side of the head. Most fish use these for stopping, steering and turning, but some swim with their pectoral fins.

4. PELVIC FINS are the pair of fins on the fish's "chest" or belly. While they may help with steering and balance, in fish that rest on the bottom the pelvic fins may be specialized for crawling or hanging on like a suction cup.

   a. Does your fish have pelvic fins?

5. The ANAL FIN is the single fin along the bottom, near the tail. It works like the dorsal fin.

6. How many fins does your fish have altogether?

7. Do you think all fish have the same number of fins? Give evidence for your answer.

G. MOUTH
Pick your fish up and take a good look into its mouth.

1. Does your fish have a tongue?

2. Does your fish have teeth?

3. Where are the teeth located within the mouth?

4. Given the location of your fish's mouth and the teeth it has, can you make a guess about where it might feed and what it might eat?

5. Put a stick or pencil into the fish's mouth. Does it come out the side of the fish's head? What would happen to water that went into the fish's mouth?

H. GILLS

Lift the flap on the side of the fish's head. Can you see sets of red things? These are the GILLS. They get oxygen from the water and lose carbon dioxide in exchange.

1. The covering over the gills is called the OPERCULUM. What purpose might the operculum serve?

2. Can you guess what makes the gills red?

3. What is the path of water as it moves over the gills?

4. GILL RAKERS are projections on the gills that keep the food in the fish's mouth. Why might these be needed?

II. Interesting insides

Label the parts diagram on the last page of this data sheet.

Now that you have studied the outside, it is time to open your fish up!

Find the VENT (anus) on the underside of the fish near the tail and carefully insert your scissors into the hole. Make a shallow cut from the vent all the way to the "chin" area of the fish, being careful not to disturb or cut any organs below. Gently pull apart the flaps of skin and muscle to show the organs below.

A. Blood and heart

1. Gills

Cut away the gill cover (OPERCULUM) on one side of the fish's head and carefully remove a section of the gills.

a. Look at the section of gills. Use a hand lens if you have one. Do you see the feathery GILL FILAMENTS attached to stiff GILL BARS? Cut a piece of the gill off and put it in water so you can see how feathery the gills are. This makes lots of surface area for oxygen/carbon dioxide exchange.

b. Blood in the gill filaments receives oxygen from the water. This oxygen-rich blood is pumped to all of the fish's body by the HEART, a small, dark, slightly triangular organ just below the gills. Label it in the drawing. Unlike your own heart which has four chambers, the fish's heart generally has just two. Blood flows from the heart to the gills and then directly to the body where oxygen is used and carbon dioxide is made. The blood carrying carbon dioxide goes back to the heart.

B. The food factory

There are a number of parts involved in the digestion of food in fish.

1. Behind the heart are several large, reddish organs. They are the LIVER. Find it in your fish and label it on the diagram. It helps in digestion and stores fats and blood sugars.

2. Search for a small, green-colored section of the liver. The GALL BLADDER also aids digestion.

3. Under these you should be able to find the STOMACH which is connected to the mouth and receives the food, and the INTESTINES where digestion is finished and the food absorbed. They are connected to the vent where solid wastes are eliminated.

4. Take out the stomach and cut it open; look at the contents. What did your fish eat before it was caught?

C. Sink or float

Between the stomach and the backbone is a whitish or silvery sack that may be filled with air, called the SWIM BLADDER. It is fragile and you may have deflated it already. In some fish it is connected to the fish's throat, and the fish fills it by swallowing air. In most kinds of fish, it is a sealed sack, and the gas comes from the fish's blood.

1. Describe what you think would happen to a fish that added more gas to its swim bladder.

D. Other goodies

1. Can you find two long string-like organs running along each side of the backbone? They will probably be dark in color and very thin. These are the KIDNEYS; they remove wastes from the blood. They are hard to see.

2. You may get a fish with a big sack of little yellow round things inside. These are eggs or ROE. Your fish was reproducing when it was caught.

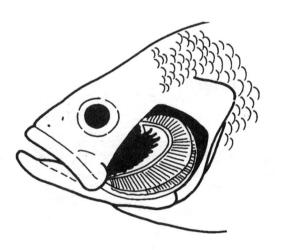

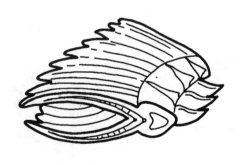

Activity 34
Sink or swim

Name _**possible answers**_

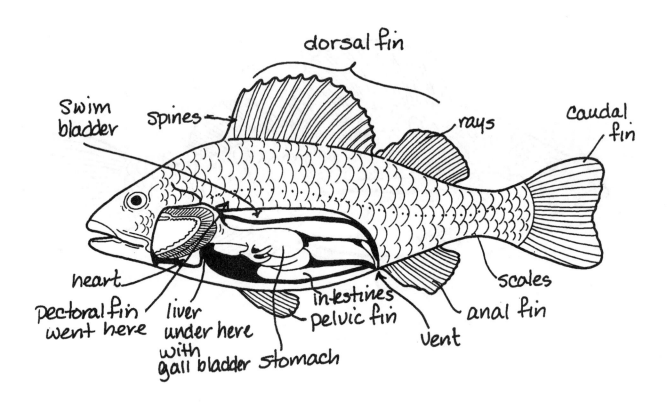

dorsal fin

Swim bladder

spines

rays

caudal fin

heart

Pectoral fin went here

liver under here with gall bladder

stomach

intestines

pelvic fin

vent

scales

anal fin

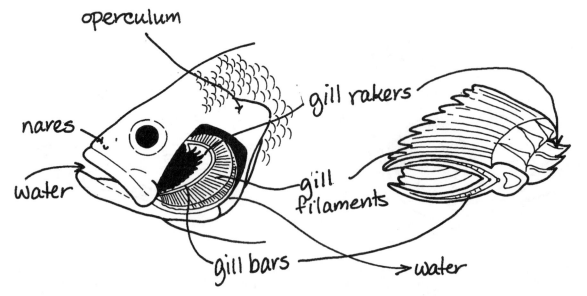

operculum

gill rakers

nares

water

gill filaments

gill bars

water

National Aquarium in Baltimore

ACTIVITY 35　Grace under pressure

How does water pressure vary with depth? Student experiments with water pressure.

Teacher's information

This is a new twist on an old exercise. One author remembers punching nail holes in tin cans as a child. The difference between this and many similar ones on water pressure and depth is threefold: 1) students collect numerical data rather than just observe the results, 2) students compare jugs with different cross sectional areas and 3) students compare the differences in size of nail holes.

Introduction

If you have children who swim, ask what happens to your ears when you dive to the deepest part of a swimming pool. Where does this sense of pressure come from? The water is pushing on their ears. What have they seen in movies that relates to water pressure? Most students will have been exposed to some undersea disaster movie. Hold up a milk jug and ask the students if they think the greatest water pressure would be near the top or the bottom. How could they test this? Do the students have a feel for the term pressure? Fill a large syringe with water. Ask the students what you have to do to make it squirt. Demonstrate with a gentle push. What do you have to do to make it squirt farther? Push hard. Can they get the relationship between how far the water went and how hard you pushed? They need to understand that the distance the water traveled was a measure of the force you used on the plunger. They can use the same relationship to measure water pressure. Now how could they apply this to measuring water pressure in the jug at different depths?

Action

One way to test their hypothesis is by measuring the differences in length of streams of water flowing simultaneously from the top, center, and bottom. Would they expect the lengths of their water streams to remain constant or to change with time? Use the smaller of their two containers to test this question.

Science skills

observing
measuring
experimenting
communicating

Concepts

Water pressure increases with water depth.

Water pressure pushes in all directions on objects in the water.

Skills practiced

measuring
averaging
graphing

Time

1–2 class periods

Mode of instruction

teacher directed group work

Sample objectives

Students develop evidence that pressure increases with water depth.

Materials

for class

water

waterproof marker to mark water level

masking tape

for each group

flat large pan (aluminum turkey
  roaster or plastic cat litter box)

1/2 gallon plastic milk jug

1 gallon plastic milk jug

measuring cup

ruler

3 thin, smooth nails of same size
  (finishing nails)

something about 4 inches high (like
  inverted measuring cup) to set
  milk jug on; must be identical for
  all groups

for each student

worksheet

graph paper

optional for the class

large syringe without needle

a large trash can 2 ft deep filled with
  water, such as custodial can with
  wheels

several small balloons, slightly
  inflated

Let them experiment to find the answer to this question. To collect consistent data they need a constant water level. This is achieved by one student gently pouring water in the side away from the holes as it drains out, keeping it even with the line you drew. The set up looks like this:

Results are improved by placing the milk jug on something to get better elevation. Whatever is used, **all** groups must have the same thing, the same height of about 4 inches! Students should record their observations as they go.

Now that the students have practiced, collect data. Start with the 1/2 gallon jugs. Leave the nails in the holes until after the carton is filled to the line. Tape the ruler to the pan to keep it from floating away. When ready to begin measuring, remove all nails at the same time and pour water in the top to keep the jug filled to the line. Measure the distance from the container that the water flows. The farther the water squirts, the greater the force pushing it out the hole. What assumption are they making about pressure? It pushes the water out farther if the pressure is greater. Think eye droppers, squirt guns, and hoses. If the distance the water squirts is a measure of pressure, then where is the pressure greater: deeper or shallower water? Do they get a continuous relationship?

Now, would it make a difference if the volume of water were greater with the same depth and nail hole size? Use the gallon jugs and compare what happens when the depth is the same and the volume is double. Make sure the nail holes within the group are identical sizes! Students are surprised to find that

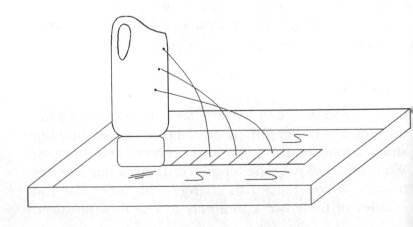

only depth is important. They have the misconception that the width and length of a body of water are important. In fact water pressure increases with depth whether you are measuring it in a pond, a lake or the ocean. The readings for gallon and half gallon should be the same if the nail holes are the same size and the distances are identical.

What about the size of the holes? Have the groups having Sets A and B compare their results. Little holes have more resistance to flow than big ones. Which squirt farther under the same pressure?

Results and reflection

Compile the class data on the blackboard. Compare the results from the different groups. Students should be surprised to learn that hole height (depth) and hole diameter are important, but that the volume of the container is not. Up to a point smaller holes have greater resistance to flow and will reduce the force

container size	Set A	Set B
1/2 gallon jugs	small nail holes	larger nail holes
gallon jugs	small nail holes	larger nail holes

of the water moving through them. Average the data and draw a bar graph to show the length of the water stream from each different hole.

Have the class discuss its results and generate some rules about depth and pressure. Under what circumstances might these rules be important for humans? SCUBA diving or going down in a submarine are two good examples.

Conclusions

Water pressure increases with depth. That is why the stream of water flows farther out at the bottom of the milk carton than at the top. The water pressure provides the push. Pressure increases by one atmosphere for every 33 ft in depth.

Preparation

Students are given two sets of variables: nail diameter between groups and container size within each group. You need to prepare the milk jugs the first time you do this. They can be stored and reused. Cut an opening in front of the handle in each with sturdy scissors. Mark an upper water level on each that is an identical height from the bottom on each below the level of the opening. See the illustration for the pattern of holes you are going to melt in the plastic. The lowest hole should be 2 inches above the bottom and the other two should be equal distances apart with the upper hole at least an inch below the water mark. Offset the holes by about 1 inch as you go down so that the streams of water do not hit each other. **All the holes on all the containers should be the same vertical distance apart.** Make the holes by heating a nail of the size the students will be using as plugs on the stove or in a candle flame, holding it with pliers. Nails referred to by carpenters as finishing nails make the nicest holes. A small finishing nail is 1 to 1 1/2 inches long. A larger finishing nail would be 2–3 inches. Carefully melt the holes where marked. Do not make the holes larger than the nails as they are going to serve as plugs when the jug is full. Make two sets of jugs, using the table. A group either gets Set A or Set B. Half the groups get Set A jugs, and the other half, Set B jugs.

Outline

before class

1. prepare jugs

2. copy worksheets and graph paper

during class

1. discuss student perceptions about water depth and pressure from personal experience and from disaster movies

2. state question about pressure versus depth and discuss how it could be measured

3. demonstrate pressure versus distance squirted with a large syringe

4. students use 1/2 gal milk jugs and measure distance water squirts from three holes

5. compare with three holes at same depth in 1 gal jug

6. compile results on the board and compare effect of size of nail holes

7. make data table and graph

Extension and applications

1. Prepare to get wet. A separate question is how water pressure affects gases trapped in animals or submersibles that descend into the water. Introduce a challenge, to be attempted in teams of no more than two at a time: can the students observe (feel) a difference in the volume of air in the balloon as it descends to the bottom of a trash can full of water? This is a wet activity! Students should roll up their sleeves and reach down only as far as they can comfortably. They should try to keep their hand in the same position around the balloon as they move down into the barrel. They should be able to feel the balloon shrink in their hand. Another method is to squeeze the balloon at the surface and then observe how much they can ease up on their squeeze as the balloon descends. Can they devise a way to measure the change in size? Establish rules in advance concerning the whereabouts of the water and the balloons.

   Compare balloons filled with air and with water. Does the water-filled balloon change with depth? No. Unlike gases, water is not compressible, so any animal that does not have gas-filled spaces like lungs or a swim bladder does not get squashed by pressure. Some fish migrate from 1000 m to the surface and back down each night with no problem. So do many large zooplankton.

2. Invite a scuba diver to speak to the class and expand on the pressure/depth relationship and its importance to people using self-contained underwater breathing apparatus (SCUBA). If you do not happen to know a diver, you might be able to enlist the help of someone who works at a local dive shop or YMCA.

Activity 35
Grace under pressure

Name _____

1. List the questions you are trying to answer with this experiment.

2. Share class results on the board. Discuss how to average them. Then make a table showing the average for all data for the large and small jugs, with large and small nails. Show your units.

	distance water flowed out from the jug			
location of hole	small nail holes		large nail holes	
	1/2 gal jugs	1 gal jugs	1/2 gal jugs	1 gal jugs
top				
bottom				
middle				

3. Use graph paper to make a bar graph showing these data graphically. Label it completely. Attach it to this worksheet.

4. Where was the water pressure greatest? Give evidence for your answer.

5. There were several variables in this experiment: size of holes, height of holes and volume of the containers. Which were important? Give evidence for your answer.

Activity 35
Grace under pressure

Name ___possible answers___

1. List the questions you are trying to answer with this experiment.

1. What is the relationship between how deep the water is and how strong the pressure is? Does the water push harder at greater depth?

2. Does it make any difference if the container is bigger if the depth of the water is the same?

3. Does the size of the hole the water comes out of make a difference in how far the water squirts?

2. Share class results on the board. Discuss how to average them. Then make a table showing the average for all data for the large and small jugs, with large and small nails. Show your units.

	distance water flowed out from the jug			
location of hole	small nail holes		large nail holes	
	1/2 gal jugs	1 gal jugs	1/2 gal jugs	1 gal jugs
top	2.1	2.2	2.4	2.3
bottom	4.8	4.7	5.3	5.2
middle	7.6	7.2	8.6	8.4

3. Use graph paper to make a bar graph showing these data graphically. Label it completely. Attach it to this worksheet.

4. Where was the water pressure greatest? Give evidence for your answer.

If you assume that the greater the pressure, the farther the water squirted, then the pressure was greatest at the deepest part of the jug because the water went farther out of the bottom holes.

5. There were several variables in this experiment: size of holes, height of holes and volume of the containers. Which were important? Give evidence for your answer.

The size of the holes made a little bit of difference. The water squirted a bit farther out of the big holes. The size of the container made no difference; only the depth was important. The water squirted the same distance from the big and small jugs. The depth was the important variable. It gave the big differences.

ACTIVITY 36 At the races!

How do fish swim? What are the correlations between body shape, swimming technique and speed? Measurements and observations made on a trip to an aquarium, zoo or science center which displays fish.

Teacher's information

Fish have a number of physical adaptations for movement in water. They must overcome drag and have a means of propulsion and maneuverability in order to move through water efficiently. A variety of body shapes and structures meet this goal. Not all fish are swift and agile, however. Losses in swimming efficiency in certain species are offset by other adaptations such as camouflage, bony plates or behaviors. All adaptations for movement reflect an animal's "lifestyle" which is in turn influenced by the physical factors of its particular environment.

This activity is designed to be done at an aquarium, science center or zoo which displays a variety of fish species. It works particularly well with bony fish or sharks if they are exhibited in such a way that they can swim continuously in one direction. Unlike most activities done at such sites, *At the Races* collects numerical data.

Introduction

Have students read *Moving through Water* as homework. Review the terms used in it. Discuss the research project for the field trip so that students know what they are going to be doing ahead of time. Pick the exhibit in which you will be working. It should have fish which swim freely and should be large enough for the students to work around it. For example, at the National Aquarium in Baltimore, both the Open Ocean (sharks) and Atlantic Coral Reef are large exhibits which allow the fish to swim in giant circles. Two windows and one intervening mullion in these exhibits equal 11 feet in distance. Students time how long fish take to swim from the edge of one window to the opposite edge of a second window.

Science skills
observing
measuring
organizing
inferring

Concepts
Fish are adapted to move in water in many different ways.

There is a correlation between the body shape and swimming speed of different fish.

Skills practiced
measuring time and distance
calculating average speeds

Time
1–2 class periods plus field trip

Mode of instruction
independent group work

Sample objectives
Students calculate rate of movement (speed).

Students correlate fish body shape and swimming technique with swimming speed.

Builds on
Activity 29
Activity 31
Activity 34

Materials

for each student
copy of *Moving through Water*
for each pair
clipboard with pencil tied on
data sheet
stopwatch or watch that measures
    seconds
for the class
fish display at aquarium, zoo or
    science center
tape measure or carpenter's rule

Preparation

Schedule the field trip. Actually, this needs to be done at the beginning of the school year. Avoid busy times like late spring. In picking a date, remember that there are always fewer people on Mondays and the most on Fridays.

Outline

before class
1. schedule field trip and bus

2. copy data sheets

day before field trip
1. review fish anatomy

2. hand out homework reading
Moving through Water

on field trip
1. pick fish, identify, observe and race

2. record results

back at school
1. compile data

2. compare observations

Action

Hand each team of two students one clipboard, data sheet, pencil, and a stop watch or a watch with a second hand. Determine the "race course" and measure and record its distance. The race course might extend from one end of the tank to another, or if it's a very large exhibit, the width of one or two viewing panels. Each team should choose two species of fish to time. Try to get a variety of fish among the teams. Have them watch the fish for several minutes to make sure the fish they choose swim relatively straight.

Before conducting the "trials," the team must identify the fish, placing their names in the spaces under "Fish #1" and "Fish #2" on the worksheet. If the facility does not have an identification label system, make arrangements for a docent or staff member to briefly help your students with identification if you are not comfortable with this. Even more important, record each fish's body shape and the fins it uses for swimming.

Now for the races! Let the students look at each other's fish and "bet" or predict which they might think is faster. Next the students will time each fish as it swims its "course," recording three trial times. Then let the students enjoy the rest of their field trip. The next day, back at school, have the students compute the average swimming speed for each of the two species for each team as distance traveled per unit time (i.e., feet per second or meters per minute). Rank the fish in terms of fastest to slowest on the board.

Results and reflection

Have the students discuss their observations and measurements. Which was the fastest fish? Does the speed correlate with fins used, kinds of tail or body shape? Fish that are streamlined and swim using forked caudal fins are the fastest fish. However, some streamlined species with less strongly forked broad tails will be observed to move very slowly, such as tarpon. These fish do not move quickly unless they have to. Then they are capable of short bursts of very fast speed. These fish have a broad base on their tail. What are the other means of propulsion the students observed? Which are the slowest fish? Depressed fish and those that move with their dorsal and anal or pectoral fins are not extremely fast in terms of forward movement, but can dart and turn swiftly, and maneuver in tight spaces.

As students determine the relative speeds of the fish tested, have them look for other adaptations that may offset an animal's deficiencies of speed. How does it protect itself? Where in the habitat does it live? What food might it be adapted for catching if it can't go after swiftly moving prey?

Conclusions

Fish have different methods for moving through water. Some methods are better than others. Speed is not the only thing that is important in terms of survival. A fish that is not speedy has other useful adaptations.

Using your classroom aquarium

You can hold fish races in your own tank if it is a big one. Most likely the aquarium fish you have chosen will not be fast swimmers. Have the students analyze ways that your fish move and which fins they use. If you have any invertebrates in your aquarium, compare their swimming habits with those of fish. There are not too many invertebrates which have mastered swimming. The jet-propelled squid is one. Most move over surfaces or drift with the currents.

Extension and applications

1. Build a Fish. After visiting the aquarium or science center, have each student design a fish. After naming the new species, have each student write a label describing the fish's habitat and its behavior. Then build the fish whose body shape and structures would enable it to survive in the habitat described in the card. Some suggestions for materials to be used include:

fruits and vegetables	clay
pinecones	toothpicks
wood scraps	buttons
wire	construction paper
sequins	markers
aluminum foil	styrofoam
tissue paper	pipe cleaners
glue	papier mache around balloons

Moving through Water: How Fish Swim

Water is harder to move through than air. It resists movement more. Think of how hard it is to run in a swimming pool. Fish must deal with this problem. There are many different solutions. Fish have special body shapes for different ways of swimming and use their fins in special ways. Here are some basic fish shapes and ways of using fins for swimming. The words that may be new for you are written in capital letters.

Body shapes

Some body shapes slip through water easily. The FUSIFORM or STREAMLINED body shape is extremely efficient for moving through water. Most constantly swimming, fast-moving fish have this torpedo shape. They usually swim in open water away from obstructions like a coral reef or kelp forest (giant seaweed). Fish that swim in fast-moving water in streams and rivers also have this shape. In this case, the fish swims into the current just to stay in place. Ocean examples include tuna, mackerel, jacks, bass, striped bass and oceangoing sharks while trout and pike might be found in a river. Fast boats and submarines copy this fish design.

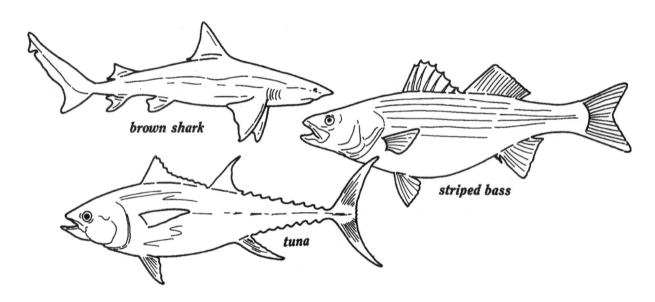

brown shark

striped bass

tuna

LATERALLY COMPRESSED fish are flattened from side to side. They look thin when viewed from the front. They are not as swift as fusiform fish, but still slide through the water well. Many laterally compressed fish live in relatively quiet waters such as ponds, lakes, or coral reefs, or they are schooling fish found in shallow open waters. This body shape, though not as fast as the fusiform shape, is well-designed for maneuvering in dense cover and making short, quick turns. Examples include pumpkin-seeds, bluegills, sergeant majors, butterfly-fish, triggerfish, pompano, and lookdowns.

lookdown

The flounder is a laterally compressed fish which is camouflaged and rests on the bottom. Both eyes are on one side of its head. One eye migrates to the other side as it is growing up. While it does not move often, a flounder can be really fast as it darts up to catch an unsuspecting small fish that swims over it.

Bottom-dwelling fish generally have a DE-PRESSED body shape which means they are flattened from top to bottom. As you might imagine from their shape and habitat, these may not be fast swimmers. While they may not swim fast, they have other adaptations such as coloration or burrowing into the substrate which help them hide from predators and catch prey. Some examples are skates, rays, sculpins, and catfish.

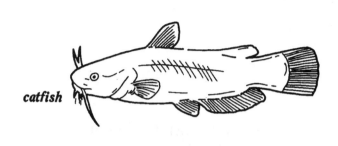

catfish

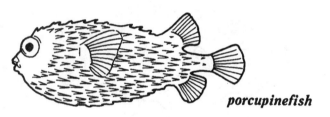

porcupinefish

Finally, some kinds of fish have cube or spherical shapes which force them to swim quite slowly. They generally have very good defenses against predators such as spines or poisons and eat plants or slow moving. Puffers, porcupinefish and trunkfish are among these slow swimmers.

Fins

Most fish depend on the CAUDAL FIN, or tail fin for sustained swimming or bursts of high speed. Bony fish with a narrow connection of tail to body tend to swim constantly, while those with a broad base on their tail swim with brief bursts of speed. Fish that swim with their tails can only swim forward. They cannot back up. Sharks generally use their caudal fin for power. One interesting thing is that the size of the shark's second dorsal fin relates to its swimming speed. The smaller the second dorsal fin, the faster the shark, but we do not know why.

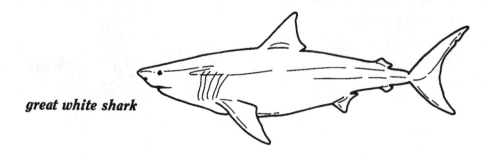

great white shark

Pairs of PECTORAL FINS and PELVIC FINS are used to steer up or down, turn or stop by fish that swim with their caudal fins. Other kinds of fish use their pectoral fins for swimming, paddling their way through the water. They are slow swimmers, but very maneuverable in tight spaces. Parrotfish and wrasses use their pectoral fins to swim.

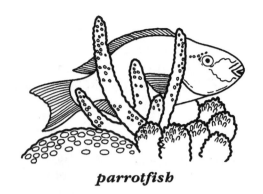

parrotfish

DORSAL FIN and ANAL FINS work like the keel on a boat, keeping a fish upright in the water instead of rolling over. Also, some fish use these fins for swimming. Waves pass down both fins in an s-shape, pushing the fish through the water. Fish that swim this way can go backward as well as forward which means they can back into caves and crevices.

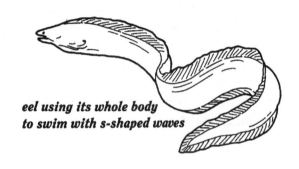

eel using its whole body to swim with s-shaped waves

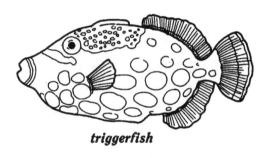

triggerfish

Seahorses are unusual. They swim upright, using DORSAL and PECTORAL FINS. This slow way of swimming helps seahorses maneuver among plants.

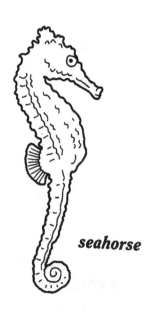

seahorse

Activity 36
At the races!

I. On your field trip

A.   Length of the race course (distance) is _____.

B.   Fish number one is a _____.

    1.   It uses its _____ fin(s) to swim.

    2.   Its body shape is _____.

    3.   Time it took this fish to swim the race course in seconds:

       a. _____ b. _____ c. _____

    4.   Average speed _____ (remember this is distance per unit time)

C.   Fish number two is a _____.

    1.   It uses its _____ fin(s) to swim.

    2.   Its body shape is _____.

    3.   Time it took this fish to swim the race course in seconds:

       a. _____ b. _____ c. _____

    4.   Average speed _____ (remember this is distance per unit time)

II. Back in the classroom

A.   Describe the body shape and swimming method used by the fastest fish measured.

B.   Describe the shape and swimming method of the slowest fish anyone found.

Activity 36
At the races!

Name   possible answers

I. On your field trip

A.   Length of the race course (distance) is ___11 feet—two windows___.

B.   Fish number one is a ___yellow tail snapper___

   1.   It uses its ___tail (forked)___ fin(s) to swim.

   2.   Its body shape is ___fusiform___.

   3.   Time it took this fish to swim the race course in seconds:

   a. __3 sec__ b. __4 sec__ c. __2 sec__

   4.   Average speed __3.67 ft/sec__ (remember this is distance per unit time)

C.   Fish number two is a ___triggerfish___.

   1.   It uses its ___dorsal and anal___ fin(s) to swim.

   2.   Its body shape is ___laterally compressed___.

   3.   Time it took this fish to swim the race course in seconds:

   a. __9 sec__ b. __11 sec__ c. __12 sec__

   4.   Average speed __1.04 ft/sec__ (remember this is distance per unit time)

II. Back in the classroom

A.   Describe the body shape and swimming method used by the fastest fish measured.

My yellow tail snapper was one of the very fastest fish. They all had very forked tails. Some were fusiform like mine, but some were laterally compressed like the lookdowns.

B.   Describe the shape and swimming method of the slowest fish anyone found.

The slowest fish was a porcupinefish that paddled with its pectoral fins and was almost a sphere. There were also some fish that didn't leave their home spot so we could not time them, like the damselfish.

SECTION V Light in Water

Teacher's information

The physics of light in water

Water absorbs light. Since light is essential to plant growth and almost all living things depend on plants either directly or indirectly for their food, light is very important in aquatic habitats. As water gets deeper, it gets darker. Even in very clear water that does not have sediment or phytoplankton (single-celled algae) blocking light penetration, it looks dark at 150 m although a bit of light penetrates to 600 m. This lack of light in deeper water has two consequences. The first is that algae or plants in the ocean or in lakes grow well only within about 100 m of the surface in very clear water. In water that is TURBID (cloudy with sediment or phytoplankton), the plants or algae must be even closer to the surface. Second, rooted plants or seaweed that must grow from the bottom up live in shallow water near shore.

Light is also necessary for vision, an important sense for many animals. Animals that depend on natural light for their vision are restricted to relatively shallow water in an ocean thousands of meters deep. Animals that are bioluminescent, making their own light, have an interesting solution to this lack of light in deep water. Animals that search for prey or that need to find others of their own species in shallow murky water or in deeper dark water make use of senses other than vision. These include adaptations for the production and reception of sound (vibrations in the water), chemical information, or electrical fields.

In addition to the general lack of light, colors (wavelengths of light) are differentially absorbed by water. Red and orange are absorbed first while blue penetrates best. For this reason, underwater movies or photographs made with natural light look blue. Those that show bright colors are made with artificial lights carried by the photographer. The brilliant colors we associate with some kinds of marine animals like sponges are invisible in their natural habitat! Other marine organisms like colorful coral reef fish live in shallow water where their colors can be seen and have a function. Color vision is common in fish that live in bright shallow water, but not in those that live in deeper water or way out at sea where many wavelengths of light are nonexistent.

Light and photosynthesis

With the exception of unique aquatic species that live near deep ocean vents or in chemical rich, oxygen poor mud, organisms that use light for PHOTOSYNTHESIS produce the food that sustains all other living things, either directly or indirectly. Things that photosynthesize are all incorrectly called "plants" by many people, but include a wide range of organisms from the familiar higher green plants to ferns and mosses to single-celled ALGAE (phytoplankton) to multicelled algae (seaweeds) to many kinds of bacteria. All of these things produce FOOD (stored energy and complex compounds) from inorganic elements and are lumped into a group called PRIMARY PRODUCERS.

Primary producers (often simply called producers) use sunlight to provide the energy for the chemical process that takes place within their cells called PHOTOSYNTHESIS. As the name implies, light (photo) energy is used to make (synthesize) some of the compounds the organism needs. The raw materials for this process are carbon dioxide and water. The oxygen that is produced is a waste product and comes from water molecules split by light energy. The food energy these organisms produce is stored in themselves—their cells, bodies, stems, leaves or seeds.

Food chains

The producers form the base or bottom of the FOOD CHAIN, also called a FOOD WEB by some because it is not as simple as a straight chain. (Ecologists generally use the term food chain.) Animals that eat the producers or other animals are called CONSUMERS. Consumers that eat primary producers directly are called HERBIVORES which comes from Latin roots meaning "grass-eating." Animals that eat other animals are called CARNIVORES from Latin for "meat-eating." Some kinds of animals eat both plants and animals and are called OMNIVORES from the Latin for "all-eating." The general term PREDATOR applies to an animal that eats another. The word PREY refers to the animal that is eaten. A given animal might be a predator on one species and the prey of another. For example, an ocean food chain might start with phytoplankton as the primary producers.

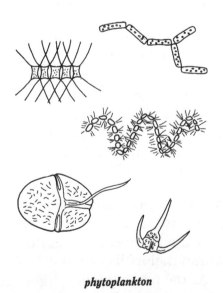

phytoplankton

These might be eaten by small zooplankton (herbivores) which would, in turn, be eaten by larger zooplankton (carnivores). The zooplankton might be eaten by small fish which are then eaten by larger fish— both carnivores. These larger fish might be preyed upon by very large fish such as tuna or big sharks—the top carnivores. This chain has six links.

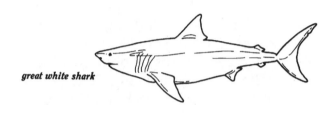

great white shark

Energy and food transfer

At each stage of the transfer of food from one level of the food chain to the next, the food and its stored energy are used for many different things. Some is used for growth or stored food (fat or oil). Some is used for reproduction. Some is wasted as undigested material lost in feces or the bodies of dead organisms. Much of the food that an animal eats is broken down in a process called RESPIRATION that takes place inside cells. In this process, oxygen combines with the carbon and hydrogen atoms in the food to make carbon dioxide (CO_2) and water (H_2O). The energy released is captured and used to do work such as movement, making new molecules or moving molecules around. Since no energy transfer is perfect, some energy is also lost as heat. Food that is broken down is said to be "burnt" since the end products, carbon dioxide and water, are the same as those of combustion. Food energy is measured in calories, so animals are referred to as burning calories when they do cellular respiration.

Food that is burnt or broken down is gone from the food chain and not available to the animals at the next level. Consequently, each level of the food chain above the level of the producers has less available food than the level below it. The food available to the next level of the food chain is in the bodies of the plants or animals at the lower level and is called the STANDING CROP. The amount of food available at each level may be expressed as total weight or as calories.

What determines the length of a food chain? Why do some habitats have longer chains than others? First, the amount and quality of food stored in the producers has an effect. Since energy is lost at each stage, if you start with a great deal of food or with very high quality food that has lots of nutrition, there will be more left at the higher levels. Second, efficient use and transfer of energy makes for longer food chains. If the animals at one level are very inefficient and burn most of the energy they consume, they will not be making much of themselves for the next level to eat. In general, food chains in aquatic habitats are often rather long when compared with those on land, often having six or seven links in the ocean.

Human impacts on ecosystems and food chains

Human activities may affect food chains in many ways. Pollution and habitat destruction cause obvious changes that are easy to see. Dead fish are hard to ignore. It is also easy to understand the consequences of filling a pond or marsh in order to build a shopping center or to add another field to a farm. The whole habitat goes away. Much more subtle changes may result from the extensive harvesting or removal of specific levels of the food chain or the introduction of ALIEN species.

When we selectively remove or add fish species, we are making changes that may have a significant impact on the other animals. Very careful planning of fishery regulations are necessary to be able to remove animals from a system on a long term basis without destroying the ECOLOGICAL BALANCE of the entire system. The result is a calculation of how many fish can be taken year after year without upsetting this balance and is called the SUSTAINABLE YIELD. People who study these problems are in a field called FISHERIES MANAGEMENT. They set fishing limits for both sport fishing and commercial fishing and enforce them. Most of the major marine fisheries under United States control are not currently being managed for sustained yield because politics, not science, determines the regulations.

Adding ALIEN SPECIES to an ecosystem can also have a profound and unexpected outcome. They can out-compete native species or they may be too effective a predator on native species. Sometimes the additions are accidental, such as the zebra mussels which are spreading in fresh water all over the eastern United States. Sometimes the additions are done on purpose such as the carp that now live in most slow moving freshwater habitats.

Extinction

The loss of species from ecosystems reduces the BIOLOGICAL DIVERSITY or biodiversity of these systems. Species may go EXTINCT. The causes of extinction were discussed in the preceding sections: habitat loss, pollution, over-harvesting and the introduction of alien species. An example of current concern is the precarious condition of salmon in many Pacific Northwest rivers. Habitat loss or alteration due to deforestation, hydroelectric dams, agriculture and alien salmon in the form of hatchery reared salmon which are genetically different than the wild salmon (all human activities) have pushed many races of salmon to the brink of extinction.

Modeling and games: learning how systems work

The concepts of food chains can be communicated to students by letting them be the animals in a simple food chain and playing a game to see who survives. The rules of the game are the rules that govern the survival of plants and animals in the wild. When they have mastered the basic rules that govern food chains, then the students can experiment with the system by changing one rule at a time (remind them—only one variable per experiment). They can ask "what if" questions, predict the answers and then run an experiment to see what happens.

This kind of SIMULATION of natural systems is a form of MODELING. Sophisticated modeling can be done with computers. Educational computer software that allows students to continue to ask "what if" questions about food chains is available from educational catalogs. Because the computer runs the simulation much faster, students can collect a great deal of "data" using a computer simulation. Such experiments might make a very nice science fair project. More sophisticated programs for high school students allow them to develop the models themselves. Students can also invent their own games which are models of biological processes.

ACTIVITY 37 Light to sea by

What happens to light when it moves through water? What happens to light in deep water? Student observations.

Teacher's information

This activity asks questions about both changes in light INTENSITY (the quantity) and QUALITY (the colors present) when light shines through water. The results are not clear-cut measurements, but value judgments (observations) made by students who may have different sensitivity to colors. This is a good time to discuss the use of instruments to make measurements that are standardized and do not include as great a possibility for error as human observation. If you have computer probes that measure light, you can demonstrate their use in this lab or have the students use them and compare results with student perceptions. Variability in perception means that there may be more than one answer. That is why scientists go to great lengths to avoid value judgments and to use standardized measurement.

Introduction

Different kinds of introduction for this exercise are appropriate, depending on the background of your students. Have they ever looked at a spectrum? A rainbow? Do they understand that light bouncing off of an object is reflected to their eyes? Do they know that the colors of objects are determined by which colors (wavelengths) of light they reflect? Begin with a discussion of what your students know about light. Make sure they remember the difference between quantity or brightness (INTENSITY) and QUALITY or colors of light.

Action

How does water affect sunlight quantity and quality available to plants and animals living in water? Show the jars of clean and dirty water. Put white paper behind them. Tell the students you are going to shine a light through the jars. Ask the students to predict which transmitted light will be brighter: the light shining through the clear water or the dirty (TURBID) water? Which jar transmitted a greater quantity

Science skills

observing
inferring
predicting

Concepts

Water absorbs light.

Water absorbs different wavelengths at different rates.

Things like sediment suspended in the water absorb light.

Time

1–2 class periods

Mode of instruction

teacher demonstration
teacher directed group work

Sample objectives

Students observe the effect of water on light transmitted through it.

Students observe that substances in water affect light.

Materials

for class

underwater pictures from scuba and
    natural history magazines
construction paper
prism
movie screen or sheet of white
    cardboard or white wall
thin cardboard cut to fit projector
    light source with slit (cereal box
    weight cardboard)
light source—air-cooled slide or
    filmstrip projector, strong
    flashlight, sun beam
clear qt jar of clean water
clear qt jar of dirty water (add 1/2
    cup of fine particle dirt and shake)
blue colored cellophane (and red and
green if you have it)

of light? Can they identify a natural event that would mimic the turbid jar? Remind them of work done in earlier activities.

Then demonstrate light quality. Use a prism and the slide projector or a beam of sunlight in a darkened room to show the colors that make up white light. Make your classroom as dark as possible so the students can see this clearly. What are the colors? What is their order in the spectrum? The order is in increasing wavelength from blue (short) to green to yellow to orange to red (long). Have the students write the colors out in order as they observe them.

What happens if you put colored cellophane over the light source (the end of the projector lens or the flashlight) before it reaches the prism? Try it with several layers of blue and then red and green if you have them. The students should be able to see that parts of the spectrum are blocked by the transparent cellophane. It is absorbing some of the colors of light. The color transmitted is the color of the material. Put increasing numbers of layers of blue over the lens. Do not discuss this.

Then ask the groups of students to address two questions. Does water absorb light, reducing the quantity of light in deeper water? Does water absorb some colors of light more than others? Give groups of students pictures taken underwater under natural light (everything looks blue), close-ups with bright colors and pictures with a bright foreground and blue background. Can these differences in underwater pictures be explained? Let groups discuss their ideas for about 5–10 minutes. Have each group write its ideas and answers for the two questions. Have the class discuss and clarify its ideas. Pass around one or more pictures of divers using a camera with a flash or other light source. Does this help explain what the students have observed in the set of pictures? Compare the blue pictures with what happened when you put blue cellophane over the light. Blue is transmitted and the other colors are absorbed in the same way that water absorbs some wavelengths and transmits blue.

Results and reflection

Quantity of light in water: The photographs showing blue look dark. The quantity of light in deeper water is reduced just as sediment in the water reduced the quantity of light transmitted through the jars of water. Ask the students to give evidence for this from their observations.

Quality of light in water: Students should be able to state that some of the colors of light (some wavelengths) are absorbed more than others. Blues (short wavelength and high energy) are absorbed least. Reds, at the other end of the spectrum, are the longest wavelength and the lowest energy. They are absorbed quickest. Ask them to give evidence for this from their observations.

How do these results relate to marine and aquatic environments? Students should be able to conclude that the deeper you go in a big lake or in the ocean, the darker it gets. They should also be able to understand that suspended sediment or particles in water will block light, even in very shallow water. It is a bit harder to understand that as you go deeper, things have different colors than they do in shallow water or when a diving light is used. Illustrate with the photographs taken under water. Those that are taken with natural light look very blue because water absorbs the other colors of light. Those taken with a flash show colors of things that are never seen under natural conditions because the wavelengths of those colors do not exist underwater. The only way the colors show is if the wavelengths that give bright colors are brought down with the diver. In subsequent exercises, you can explore the implications of these conclusions for the plants and animals that live in aquatic and marine systems.

Conclusions

Both the quantity and quality of light available to plants and animals living in water are very different than what is experienced on land. Water absorbs light, and it absorbs the different wavelengths (colors) at different rates. In very deep water there is no sunlight.

Preparation

Prepare the jars of clean and dirty water. Prepare the pictures and light source. You only need do this once as the items are permanent. Sort pictures taken underwater into 4 groups: wide angle pictures in which the dominant color is blue, pictures taken close up with bright colors, wide angle pictures in which the foreground has color and the background is blue (typically a diver in the foreground), and pictures with an underwater photographer showing the camera and flash. Your goal is to have one of each kind of picture for each group of 3 or 4. If you have only one or two showing the photographer, you can share one with all the groups. Glue the pictures to construction paper. If you can, laminate them for durability.

Prepare your light source. If you have a projector, make a cardboard square the size of a slide or part of a filmstrip and cut a thin slit in the middle about 1 in long and 1/16 in wide with a mat knife or razor blade. This allows you to create a nice spectrum when a beam of light is directed through the prism. Or make a round cardboard with the same size slit to cover your flashlight. If you are using a beam of sunlight, such as comes through old window shades, use a whole piece of cardboard with the same size slit in the middle. Practice using a prism and beam of light to make a spectrum of all the colors of light on a white surface.

Outline

before class

1. prepare underwater picture cards

2. prepare cardboard with slit

3. prepare dirty water

during class

1. review light and vision

2. compare light shining through jars of clean and dirty water

3. use a prism to demonstrate the color spectrum

4. use blue cellophane to show what happens when light shines through water

5. have students compare underwater pictures for light quantity and quality

6. discuss observations

Extension and application

1. Some animals that live in the deep ocean communicate with light they make themselves since there is not any natural light. This light is referred to as bioluminescence because it is light made by a living thing. Fireflies are another example of the use of "living light." If you have a flashlight for each student, divide them into "species" pairs and see if they can figure out how to find their mate in a dark room, using only a flashlight to communicate. They will have to work out coded sequences to go with their "species." In the wild, different patterns of many lights on one animal would also work to distinguish different species in the dark.

2. The Monterey Bay Aquarium's education staff (www.mbayaq.org) designed a great activity with lanternfish. These fish live in water as deep as 1000 m or more. Students get enlarged drawings of several species of lanternfish, which have different patterns of bioluminescent lights down their sides. Each student makes a fish and uses fluorescent paint from a craft store for the spots. The fish are mounted on sticks so the students can hold them up to be seen. Students in a dark classroom must find all the members of their species, demonstrating how lanternfish find mates in the deep sea.

bioluminescent deep sea animals

ACTIVITY 38 Hide and seek

What do animals see under water? Is camouflage the same below water as above? Students model animal behavior.

Teacher's information

As students learned in Activity 37 Light to sea by, some colors of light (wavelengths) are absorbed faster than others when passing through water, particularly the red and yellow wavelengths. Blues are transmitted best. These facts have interesting consequences for colors, color patterns, and their distribution among animals that live in water. Fish that live in shallow, well-lighted water may have color vision. But what do most fish see? Fish that live in the open ocean are generally colorblind, seeing things in shades of gray. Fish that live in murky or muddy water may be almost blind and depend on touch or electrical fields to sense their surroundings. In this exercise your students will experience what the world looks like to fish that live far enough below the surface that the world looks blue, the only color to effectively penetrate very deep.

Introduction

Ask the students to write their prediction for the answer to this question: what do you think is the use of red color in fish that live below 10 m (33 ft) in the ocean? Red color is typical of many California saltwater fish that hang out around rocks in 10 m (33 ft) or more of water. Many shallow water nocturnal fish are also red, and red is a very common color for deep sea animals in general. Can the students test this question?

Action

First the students need red fish. Have students review their knowledge of fish anatomy from Activity 34 by drawing with light pencil and cutting out a fish made of red construction paper. Did they remember paired pectoral and pelvic fins, the tail (caudal), dorsal and anal fins? Sign their name lightly on the side they drew on.

Science skills
observing
predicting

Concepts
Color patterns that are easy to see in air may be very well camouflaged under water

Skills practiced
drawing
cutting with scissors

Time
1–2 class periods

Mode of instruction
individual student work

Sample objectives
Students explore and explain why color patterns that are easy to see in air may be hard to see under water.

Students experience the problems predators face when searching for camouflaged prey and develop foraging strategies for these prey.

Builds on
Activity 34
Activity 37

Materials

for each student

8 inch square piece of blue
  cellophane (from art supply store
  or school art supply catalog)
1 ft clear tape
2 ft string
1/2 sheet heavy construction paper
  or card or poster stock cut
  lengthwise (any color, about 4.5 in
  x 12 in)
1/2 sheet dark red construction paper
  cut crosswise (6 in by 9 in)
pencil
copy of goggles template included
  here

for each group

several pairs of scissors
stapler

for class

pictures from Activity 37
darkened room

Now the students need to be able to see what fish see in 10 m or more of water. Either use pre-made goggles or have each student construct a pair of goggles using the provided pattern. Cut out the goggles template. Draw around it on stiff construction paper or tag board. Cut out the goggles. Then add cellophane. Inexpensive blue cellophane available in rolls from school art supply stores is folded in three or four layers to cover the eye holes on the goggles. Test ahead of time whether three or four layers gives the best result when looking at a red fish on a black background. The fish should disappear. The quality of the cellophane changes from time to time so testing it is useful. Tape the cellophane in place. Cut the string in half. Tie a fat knot in one end of each string and staple the knotted end of the strings to each end of the goggles. The strings are tied to hold the goggles on a child's head. Collect the goggles.

Explain to the students that they will use the goggles to see as fish see underwater. Do not allow students to wear the blue goggles for more than ten minutes. Wearing them for a long time may temporarily bleach some of their visual pigments selectively. This does not do permanent damage, but it can feel weird to have your color vision altered temporarily. It takes much longer than 10 minutes for this to happen.

When the students are not in the classroom, distribute all the red fish around the room against solid dark backgrounds—green, blue, brown, black, ma-

genta etc. Turn the classroom lights off and create dim light. It is dark in 10 or 15 m of water. Use a loop of tape on the back side to attach the fish to bulletin boards and audio visual equipment, prop them on shelves, put them in corners on the floor. Hold a pair of goggles up to check that you are placing the fish against backgrounds with the same value.

Meet the class outside the darkened room with their goggles. When the goggles are in place, have the students enter the room and sit down. Tell them they are predators searching for red fish in 10 m of water. They are wearing the goggles because blue is the primary color of light that penetrates very far into water. Have them start searching for the fish at the same time. Time them if you want to repeat the exercise without the goggles.

Results and reflection

Stop them before all the fish are found and have them sit back down. Remove their goggles. Now can they see the fish they missed? Why were the fish hard to see? The fish reflect only red. The cellophane allowed mostly blue light through. Under water there would be no red to see so the fish look black. If you wish, repeat the exercise without the goggles to compare the time it takes to find the fish when red is visible.

Did any students develop special foraging strategies? Have them discuss them. One of the most common is to use touch, brushing one's hand across surfaces. Whiskers (barbels) serve the same purpose on fish.

How did their prediction of the function of red color compare with their observations? Red is usually thought of as WARNING COLORATION, that is warning predators that something is distasteful or even poisonous. Underwater it is not seen. A fish that appears very colorful to us in air or in a flash picture may, in fact, be very well CAMOUFLAGED from predators. The fish is hard to see because red light is missing. It is being absorbed by the water and, therefore, cannot be reflected to the predator's eyes.

Preparation

You may choose to have a parent or aide make the goggles and keep them as a classroom set, or you may have the class make them and keep them. The students could make the goggles during an art class and decorate them as fish heads with opercula and pectoral fins, etc. See Action section for detailed instructions. Do not laminate them as it makes sharp edges. Students should not wear them for more than 10 minutes at a time as it can temporarily bleach selective visual pigments, making the world a different color when they are first removed. This is not a permanent change and does not hurt the student, but can be disconcerting if the student does not understand.

Outline

before class

1. make goggles or have aide or students make them

during class

1. have students predict the value of red color to a fish

2. students make red fish

3. teacher hides red fish in plain sight while students are out of the room

4. students return and use goggles to model being underwater as they search for the red fish

5. discuss value of red color and student foraging tactics

Use the color photographs to illustrate. Any colorful underwater photograph was shot with a flash which provided all the wavelengths of light. Any photo in which the predominant color is blue shows what it really looks like under water.

Discuss how the activity was a model of animal behavior. How was it like the real world? How was it different? Why was it necessary to use a model rather than the real thing? What did the students learn by using this model?

Conclusions

You cannot make judgments about animals based on human perceptions. Fish in shallow, clear water may see things in a way that is similar to us, but fish that live in dark, murky water or deep water probably do not have color vision and may use vision very little, depending on other senses.

Extension and application

1. What about the function of red color in fish that live in shallow, clear, brightly lighted water: coral reef fish? Have the students use the library to find pictures and descriptions of coral reef fish that are red. What do they have in common? They are all nocturnal! Red is the first wavelength to disappear at night. Red fish are invisible at night because there is no red light to see them by.

2. Have the students search for other examples of red animals in the sea. There are lots, from the red of the giant Pacific octopus to the red of krill to the red of deepwater crabs.

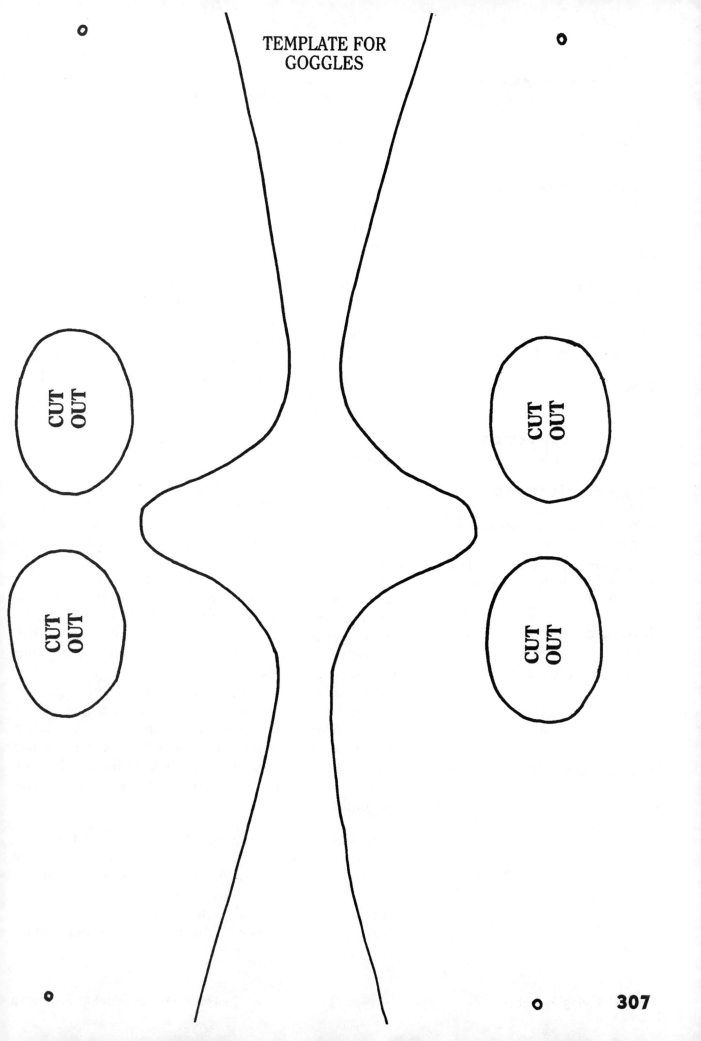

TEMPLATE FOR
GOGGLES

CUT
OUT

CUT
OUT

CUT
OUT

CUT
OUT

ACTIVITY 39 Partners in the deep

How do fish that live in deep water find others of their kind?
Students model animal behavior.

Science skills
observing
inferring
predicting
experimenting

Concepts
In deep water where natural light is very dim, animals have a number of adaptations for finding mates, including the use of sound.

Time
1 class period

Mode of instruction
whole class cooperation

Sample objectives
Students experiment with one technique, sound, that animals use to find mates.

Students discuss and analyze their strategies employed during the exercise.

Builds on
Activity 37

Teacher's information

It gets dark fast as one descends into deeper water. The light goes away, even the blue light. Sound, on the other hand, travels better through water than through air: both farther and faster (five times faster). As largely visual creatures, humans are not as accustomed to using sound as effectively as are some other animals. We have invented tools such as sonar to help us detect submarines or the contours of the bottom of the sea. Many fish and marine mammals use sound to communicate with each other. This exercise allows your students to experiment with finding others of their own kind by using sound.

Introduction

Ask each student to take about five minutes to write what he/she thinks it would be like to be in very deep water. They are to describe the physical environment around them. You may want to review what "physical" means and have them list some of the physical aspects of the environment they have studied. Then have them share their thoughts with a partner for several minutes. Finally, ask the class to give you its collective ideas and make a list on the board. At a minimum, the students should be able to tell you it gets dark, and that the light from the sun is absorbed by the water. If you have done the entire curriculum, they may also say that it is colder down deep (cold water sinks) and that the salinity is higher (just a bit saltier, but denser salty water does sink) and that the pressure is higher. Movies and TV programs reinforce the last concept frequently.

Probably no one will mention sound, but it can be noisy in the sea. Ask if any of them have ever hit the side of the bathtub while their head was under water or what it sounds like when they are swimming underwater and another person does a "cannon ball" dive. Sound travels well through water, much farther and faster than in air.

Ask them to speculate on ways the fish in deeper water find a mate. Think about what they themselves would do in a dark room to find a friend. Natural history shows and books usually emphasize light produced by living things, bioluminescence, as one way animals find each other. Students should also suggest sound.

Action

Pass out the film cans. Let them try shaking them. You can darken the room if you want or ask students to close their eyes, but this activity is hard even when you can see. Ask them to gather in groups that make the same sound by walking around and finding their own "species."

Results and reflection

When all are sure they have found their mates, have the students sit in the groups. Then one group at a time, remove the lids and see if they were accurate. Record how many were "right" and "wrong" in each group.

What strategies did they employ? Ask them to describe how they went about their search and how they thought about doing what they did. Generally, the least effective way to search is to always make noise. Going silent to listen and then "calling back" when the other individual remains silent is a better tactic. Ask the students what some of the problems were. Sound overload is one. What if a predator were listening in? What would noisy ship engines do?

Ask the students to discuss how this model was like the real thing and how it was different. Why did they have to use a model? Could they design a more realistic one? Perhaps, if your school has a swimming pool. What did they learn using a model that they could not by having someone tell them about it?

Conclusions

Sound is an effective technique for communication in water, as well as in air.

Materials
for each student
plastic 35 mm slide film canister (not clear plastic)
for the class
straight pins
paper clips
small nails
plastic bag for storage of cans

Preparation
Ask a camera store for 35 mm slide film cans or ask friends and parents for film cans. They need to be all the same kind and not clear plastic. Fill each film canister with one kind of object: 5 straight pins, or 5 paper clips or 5 small nails. Make one half the cans one type and about equal numbers of the other two kinds. If you are unable to use metal items, rice, popcorn kernels and dry beans might be substituted.

Outline
before class
1. collect film cans

2. fill with small items

during class
1. students write paragraph about being underwater

2. review student ideas and list on board

3. ask how fish in dark, deep ocean find mates

4. pass out film cans and have students form species groups using sound

5. discuss success

6. discuss nature of the model and its value

Extension and applications

1. How do fish make sounds? Have the students research the drum family of fish as well as other fish that make sounds. Names like croaker, toadfish and grunts are good clues. The swim bladder they saw in fish dissection is both a sender and a receiver of sound. Try this demonstration. Borrow two of the band's big marching drums and place them opposite each other. Hit one while students lightly touch the surface of the other. One sends vibrations and the other receives them. The swim bladder both sends and receives sound.

2. Assign research papers on all the ways fish and other aquatic and marine animals communicate with each other. Bioluminescence, sound, chemical attractants and electrical currents are among the techniques they may uncover.

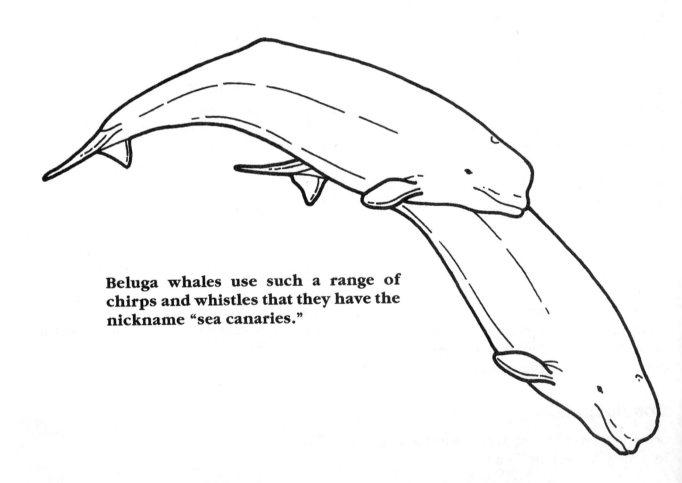

Beluga whales use such a range of chirps and whistles that they have the nickname "sea canaries."

ACTIVITY 40 A light snack

What is the relationship between light availability and photosynthesis in aquatic plants? Student experiments measuring dissolved oxygen.

Teacher's information

This experiment tests the effect of two different light intensities on the amount of photosynthesis done by aquatic plants. It shows that the amount of light available may limit photosynthesis. This concept is important to understanding several ecological conditions. The first is that the production of food by plants and phytoplankton is limited to the surface waters of lakes and oceans because light is absorbed by water. Although vast in size, oceans and large lakes are more limited in their ability to produce food than it might seem because photosynthesis requires light. Only the top layers of these bodies of water have enough light for photosynthesis. A second consideration is the effect of human actions that increase the TURBIDITY or cloudiness of water. These actions cut down on the light needed by plants to make food using photosynthesis.

Introduction

Review the results of student observations with regard to light and its absorption by water. What is the most likely effect of low light availability on the plants and algae that live in water? Introduce the concept that plants, phytoplankton and seaweeds make their own food in a biochemical process called PHOTOSYNTHESIS. The name of the process defines it: photo (light) synthesis (putting together). The energy from light is used by the plants or algae to put water and carbon dioxide together in a larger molecule. No light, no food.

Oxygen is a waste product of this process. It comes from the water molecules the plant takes apart. Can the students tell you how we could measure photosynthesis? They have already learned how to test for oxygen. They can measure how much photosynthesis takes place by measuring dissolved oxygen produced in different light conditions. Can the students think of one problem with measuring dissolved

Science skills
measuring
organizing
inferring
predicting
experimenting
communicating

Concepts
If sufficient light is available, plants produce more oxygen in photosynthesis than they use in respiration.

The amount of oxygen produced is proportional to the available light up to a point.

Skills practiced
measuring dissolved oxygen
averaging

Time
3 class periods

Mode of instruction
teacher directed group work

Sample objectives
Students design and complete an experiment to test how the amount of light affects photosynthesis rates in aquatic plants.

Builds on
Activity 14
Activity 21
Activity 22

Materials

for each large group (10–16 students)

2 pint, 12 oz or 1/2 pint canning jars with screw tops (all jars must be one size) with aquatic plants

Elodea (see Recipes): 6 strands, 40–48 inches total (pint jars) or 4 strands 20–24 inches total (1/2 pt or 12 oz jars)

2 canning jars of boiled water, sealed, at room temperature, same size as above (see Recipes)

4 dissolved oxygen sample bottles A (see Recipes)

4 dissolved oxygen test vials B (see Recipes)

4 dissolved oxygen test syringes or titrator (see Recipes)

4 aluminum pie tins to catch spills

ruler

for the class

2 jars canned boiled water (see Recipes)

two light conditions, one low and one bright, at same temperature

aged tap water

labels

pencils or waterproof markers

2 dissolved oxygen test kits

for each student

worksheet

goggles

optional

2–3 ft cheesecloth

oxygen? What did they learn about plants and oxygen in Activity 21? Review how that experiment was done and what it proved. Plants use oxygen also. The students can only measure the amount of oxygen produced by photosynthesis that is in excess of the amount being used by respiration.

Can the students help you design an experiment to test the effect of quantity of light on photosynthesis? What are the variables that must be controlled? Can the students identify an environmental variable that is often encountered with sunlight that must be controlled? Since this is a chemical process like respiration, they must control the temperature. Heat is a form of energy that often accompanies the generation of light. Ask the students to think about sitting in the sun or how hot an incandescent light bulb gets. Also, how are they going to be sure that the water itself is not affected by the light? They need controls with no plants. These are the canned water which was boiled so it will start at about the same dissolved oxygen as the plants that have been sitting overnight. Finally, what about the amount of plants in each test? They must be equal.

Show the students the equipment available to them and ask them to spend 5 minutes alone thinking and making notes about how they could use it to answer the original question. Then have them work in small groups of 3–4 to discuss and refine their ideas. Have each group create a diagram of its experimental set up, list the steps they would take and design a data table. Finally, lead a class discussion to help the students design this experiment. Ask them to justify each step. Write all the steps on the board. The step they will leave out is putting the jars in the dark overnight to make the initial dissolved oxygen lower. Review why this is necessary. To reduce the number of jars and plants needed each large team (10–16 students) will divide into 4 subunits and test one of the team's 4 jars at each step. Fill in questions 1–4 on the worksheet.

Action

Either bring out the prepared jars that have been in the dark overnight (you did the day one steps) and explain what you have done or have the students prepare them as follows. Make sure each jar is labeled.

Day one

Pour aged tap water into 2 jars. Add the *Elodea* or other aquatic plant to each jar. Measure the same total length of plant in each jar, about 20–24 inches per pint, using healthy plants. Make sure the jars are full to overflowing and seal. Place at room temperature in the dark overnight. As the students learned in Activity 21, this will remove almost all of the oxygen from the water of jars with plants. It will also saturate the water with carbon dioxide because the plants are using the oxygen in respiration and producing carbon dioxide as a waste product. If they started this experiment with the amount of oxygen normally found in room temperature water, oxygen (which is a waste product of photosynthesis) would build up and slow down photosynthesis. If you used boiled water for the plants which is low in oxygen, it would also be low in carbon dioxide which is required for photosynthesis.

Day two

Morning

The following day get the *Elodea* jars and the jars of boiled, canned water. Ask the students to review how they planned to measure how much photosynthesis is going to take place in the two light levels. Oxygen is produced as a waste product of photosynthesis, so they can measure photosynthesis by measuring an increase in dissolved oxygen. What do they need to do before putting their plants and their controls in the light? Measure the amount of oxygen

Preparation

The first time you use this with students, try it yourself first to test your light source. The experiment has three steps. How you do them depends on whether you have one set of students all day or a number of groups of students during the day. Start the experiment by putting the *Elodea* in the jars filled to overflowing with aged tap water. Put the *Elodea* jars in the dark overnight. Either you or the students may do this step which takes about 10 minutes to prepare. This step causes the dissolved oxygen in the jars containing *Elodea* to drop due to respiration. This step is essential. If you just place the plants in tap water and put them in the light, the water is already saturated with oxygen. The oxygen the plants make will come out of solution as bubbles that you cannot accurately measure.

The next day the students test each jar for dissolved oxygen and then place the jars in two different light intensities. To make sure the light is not somehow affecting the water itself, students will also open two sealed jars of boiled water, test them for dissolved oxygen. Replace all the water removed for the dissolved oxygen tests with boiled, canned water. You can use cheese cloth to make semi-shade for one set of jars and use only one light source. It is important to have a cool light source such as indirect sunlight or a fluorescent grow light that does not heat up the brightly lighted jar, introducing a second variable.

If this is done in the morning, the next step can be done in the afternoon. Four hours in indirect sunlight is enough to give good results. (The jars may sit overnight if they have a constant **very low** artificial light source.) In the afternoon students again test for dissolved oxygen to determine the effect of the two different light intensities. You may shorten the activity by setting it up yourself, putting the plants in the jars overnight and then testing the initial dissolved oxygen yourself.

at the beginning of the experiment. They can very gently pour water into the sample bottle or they may use a syringe if they do not squirt the water into the sample bottle (see Recipes for water transfer techniques). Have them do so and record the results. Replace the water lost in sampling with cool boiled tap water and reseal.

How will the students test the effect of low light on photosynthesis? Put one jar with plants and one without each in high and in low light after testing initial dissolved oxygen. One pair of jars might be in full indirect sunlight (such as a north facing window) and the other pair the same spot with 1–2 layers of cheesecloth over the jars for 3–4 hrs. Use a low artificial source of light if you are leaving them overnight. Incandescent lights produce heat, so use only fluorescent lights. Students may either put the jars two different distances from one light source (light intensity varies inversely with the square of the distance so put one set 12 inches away and the second about 17 inches away from the light to get half as much). Or students may put them all the same distance from bulbs of two different intensities such as 75 and 150 watts or put cheesecloth over one set. Avoid intense heat, but make sure the jars stay at room temperature. If 24 hrs of artificial light is used, test ahead of time to make sure the lower light intensity does not saturate the system with oxygen over that period. Plants in sealed containers stop doing photosynthesis when the oxygen becomes very high in the jar. The plants in high light might make lots of oxygen very fast and stop while the plants in the low light might keep going and finally get to the same point.

Afternoon (or next day)

At least 3–4 hrs later (indirect sunlight) or the next day with very low artificial light, test the dissolved oxygen in all the jars. Calculate the results as the difference between two measurements in each jar. Have students put their results on the board so that all the students can share the data for their worksheets. Have the students complete the worksheet.

Results and reflection

Have the students discuss the results from the table on the board. The dissolved oxygen produced should be greater in the jars that received more light.

Did the controls change? If the temperature did not change, they should have been relatively constant so the students can conclude that the source of the increased dissolved oxygen was the plants under the influence of light and not the water.

Remind them of the results from plants left in the dark in Activity 21. Considering that respiration happens all the time, the amount of actual oxygen produced is really higher than you measured because it is constantly being used in respiration as it is being produced by photosynthesis. All that can be measured in this activity is the amount left over which is called NET PHOTOSYNTHESIS. While your students may not get this analogy, it is the best one I have to offer: gross photosynthesis is like gross income and respiration is like taxes, resulting in a much smaller net income. Gross photosynthesis is assumed to be the total of net photosynthesis (which you measured in this experiment) and respiration measured under the same temperature in the dark (Activity 21). Ask the students to work in pairs to see if they can diagram the idea of gross and net photosynthesis and respiration. Then share. Here is one way: if gross photosynthesis is a big circle, then respiration would be a smaller circle inside of it and net photosynthesis is what is left inside the large circle, but outside the smaller.

If a student were an aquatic plant, where would he/she want to live and why? Near the surface or in shallow water where there would be enough light to make enough food to survive. It would be important that the water be clear so that light could get through as they learned in Activity 37.

Conclusions

The amount of photosynthesis done by water plants is directly related to the available light. Where light does not reach in deep water, no photosynthesis takes place. Only the light surface waters produce food.

Using your classroom aquarium

If your classroom aquarium receives good light and you have a good algal population or many underwater plants, you might try turning off the aeration system on two successive days. The first day leave the light on and test the dissolved oxygen level after several hours. Turn the aeration back on. If the animals did not show signs of stress, the following day turn off the aeration system, turn off the light and cover the tank with a dark cloth for two hours. Again test the dissolved oxygen. It should be lower if your plants or algae are healthy.

Discuss why plants and animals both belong in an aquarium. What is the role of the aeration system? If you carefully planned the amount of plants and animals in your aquarium, could you do without an aeration system? Theoretically, in a properly balanced aquarium with the right amount of light, the amount of excess oxygen produced in photosynthesis could balance that used by plants and animals in respiration.

Extension and applications

1. Have your students design and draw a balanced aquarium in which aeration of the water by a pump is not required. Ask them to write a paragraph explaining why they used the combination of plants and animals that they did. It is possible to find the right combination of plants and animals, but most people put too many animals in their system.

2. Have your students list all the things they can think of that make water more turbid. Some things are sediment from farms or housing developments, sewage, and industrial wastes. They might not realize that things like plant nutrients that come from fertilizer, animal manure or sewage can make the water more turbid by encouraging phytoplankton to grow too fast. The phytoplankton can actually get so dense that they keep light from reaching aquatic plants growing on the bottom.

3. For a science fair project a student might compare photosynthesis in different species of aquatic plants or she/he might use the techniques learned in Activity 40 to test the sensitivity of different species to light or to discover what level of light each species needs for photosynthesis to exactly equal respiration (the compensation point).

Activity 40
A light snack

Name _____

A. Before starting

1. What is the question that your class is trying to answer by doing this experiment?

2. Describe the experiment you and your group members are going to do. You may use diagrams or drawings.

3. List the possible variables that you are going to hold constant in this experiment.

4. What variable is going to be manipulated (what condition is being tested)?

B. Data

1. Record your large group results here:

condition	initial DO in ppm	final DO in ppm	difference
plants high light			
plants low light			
no plants high light			
no plants low light			

2. Average the differences from all of the high light samples and low light samples for the class. The averages are:

condition	difference in jars with plants ppm DO	difference in jars without plants ppm DO
high light		
low light		
difference		

C. Analysis

1. Did the light act on the water or on the plants? Give evidence for your answer.

2. Based on these results, what statement can you make about the importance of light to plants and algae in producing oxygen during photosynthesis?

3. If you were an aquatic plant, what kind of light would you want to live in? What about water clarity? Give evidence for your answer.

4. Can you figure out any other ways you might test the effect of light on a plant's ability to produce oxygen? Describe you ideas.

Activity 40
A light snack

Name __possible answers__

A. Before starting

1. What is the question that your class is trying to answer by doing this experiment?

    We are trying to figure out if increasing the amount of light will cause a plant to make more oxygen. This is a way to measure if more light will cause the plant to do more photosynthesis.

2. Describe the experiment you and your group members are going to do. You may use diagrams or drawings.

    We are going to have two kinds of jars (plants and no plants) and we are going to put them in two kinds of light conditions (high and low light). We are using jars with no plants to make sure the light is not acting on the water.

3. List the possible variables that you are going to hold constant in this experiment.

    The amount of plants in each jar is one variable. The temperature of each jar is another one. The size of the jars is also the same.

4. What variable is going to be manipulated (what condition is being tested)?

    The amount of light that each jar gets is the thing we are testing.

B. Data

1. Record your large group results here:

condition	initial DO in ppm	final DO in ppm	difference
plants high light	0.6	8.8	8.2
plants low light	1.2	5.3	4.1
no plants high light	1.4	1.0	−0.4
no plants low light	1.2	1.4	0.2

2. Average the differences from all of the high light samples and low light samples for the class. The averages are:

condition	difference in jars with plants	difference in jars without plants ppm DO
high light	7.8	0.2
low light	4.6	0.1
difference	3.2 ppm more	0.1 ppm

C. Analysis

1. Did the light act on the water or on the plants? Give evidence for your answer.

There was almost no difference between the jars with just water under the two different light conditions so the oxygen we measured in the jar with plants did not come from the water. It had to come from the plants. The plants made almost twice as much oxygen in bright light as in the lower light so in this case increasing light increased photosynthesis.

2. Based on these results, what statement can you make about the importance of light to plants and algae in producing oxygen during photosynthesis?

> Light was the thing that caused the plants to produce more oxygen which we know is a product of photosynthesis. So it would be very important for a plant to get enough light or it could not make enough food.

3. If you were an aquatic plant, what kind of light would you want to live in? What about water clarity? Give evidence for your answer.

> An aquatic plant would do best in good, clear water where the light could shine through the water. We tested light in water earlier and found that dirt blocks light from shining through water. Higher light is good because it helps the plants make food so clean water is best.

4. Can you figure out any other ways you might test the effect of light on a plant's ability to produce oxygen? Describe you ideas.

> If I had a lot of time, I could measure two sets of Elodea strands and put one set in low light and one in high light. Then I could measure how much each set grew and compare growth over a period of time like two or three weeks.

ACTIVITY 41 Muddy water

What happens to photosynthesis in plants that live in muddy water? A student designed experiment measuring dissolved oxygen in turbid systems.

Teacher's information

This experiment tests the effect of two different sediment conditions on the amount of photosynthesis done by aquatic plants. It challenges the students to test the effect of TURBIDITY or cloudiness of water. In natural ecosystems turbidity cuts down on the light needed by plants to make food using photosynthesis.

This is a repeat of Activity 40, but with the students having more control over what they do. The students design the experiment based on what they did in Activity 40. Again, the experiment has three steps. Review Activity 40. The difference is that this time instead of two light sources, there is only one. Water cloudy with sediment is the variable. This is an outline of how they might design the experiment. Start the experiment by putting the *Elodea* in the jars filled with aged tap water in the dark overnight. Having proved in Activity 40 that light does not act on the water, the water control may be deleted. Either you or the students may do this step which takes about 10 minutes to prepare. This step causes the dissolved oxygen in the jars to drop due to respiration. You must do this first step. If the plants are put in tap water in the light, the water is already saturated with oxygen, and the oxygen the plants make comes out of solution as bubbles that you cannot measure accurately.

The next day the students test each jar for dissolved oxygen, replace the lost water with boiled tap water and place the jars in light in two aquariums, one set in cloudy water and one in clear water. If this is done in the morning, the next step can be done in the afternoon or the jars may sit for as long as 24 hrs if they have a constant light source. On the third day, students again test for dissolved oxygen to see what the effect of the two different kinds of water—clear and cloudy—was. You may shorten the activity by setting it up

Science skills
measuring
organizing
inferring
predicting
experimenting
communicating

Concepts
If sufficient light is available, plants produce more oxygen in photosynthesis than they use in respiration.

The amount of oxygen produced is determined in part by available light.

One consequence of sediment suspended in water is that it may reduce photosynthesis, and thus, the growth of plants in aquatic systems.

Skills practiced
measuring dissolved oxygen
averaging

Time
2–3 class periods

Mode of instruction
teacher directed group work
independent student work

Sample objectives
Students design and complete an experiment to test how the sediment in water may affect photosynthesis rates in aquatic plants.

Builds on
Activity 16
Activity 21
Activity 40

Materials

for each large group (4–8 students)

2 pint or 1/2 pint fruit jars with screw
    tops

aquatic plant *Elodea*, about 20–24
    inches of strands per pint (see
    Recipes)

2 dissolved oxygen sample bottles A
    (see Recipes)

2 dissolved oxygen test vials B (see
    Recipes)

2 dissolved oxygen test syringes or
    titrators (see Recipes)

2 aluminum pie tins to catch spills

ruler

for the class

1–2 gal aged water

indirect sunlight or fluorescent light

2 dissolved oxygen test kits

pint boiled tap water cooled and
    sealed (see Recipes)

4–8 cups of dirt with fine particle
    size, not potting soil

waterproof labels

pencils or waterproof markers

two small (1–3 gal) plastic or glass
    aquariums

bucket

for each student

goggles

Preparation

A day ahead thoroughly mix the dirt with water, in a ratio of about 1 cup of dirt to a quart of water. Soil with fine particles (clay) is necessary to have suspended sediment particles. Commercial potting soil will not work. Make enough to fill one small aquarium. Let the large soil particles settle out and siphon off the suspended sediment in water for use in the experiment. Boil and seal a pint of water.

yourself, putting the plants in the jars overnight and then testing the dissolved oxygen yourself if you let the students design what you do. If the cloudy water is not very dirty and you leave the plants in bright light for 24 hours, the turbid water plants may do enough photosynthesis to "catch up" with the clear water plants. One good resolution to this problem for a middle school teacher is to have the first several classes of the day set up the experiment, place the aquarium in good strong light and have the last classes finish it. Then the classes have to share data. Or try it both ways, one set short and one overnight in artificial light and compare the results.

An alternate test that the students might want to do involves putting the plants directly in the muddy or clear water inside their jars and doing the same sequence of tests. This is an opportunity to discuss variables and controlling variables again.

Introduction

Review the results of Activity 40. What was the effect of low light availability on plants that live in water? Review student results and conclusions. Can the students name an environmental condition that would reduce the light available to plants living in shallow water? One such condition occurs when suspended sediment from soil erosion causes turbidity or cloudiness which blocks light. Show them the sediment filled, turbid water and challenge them to use what they learned to do in Activity 40 to test the effect of sediment on photosynthesis.

Action

Give groups time to discuss and write down the steps in their plan for the experiment and data tables. Ask them to imagine themselves doing each step as they write in order to make sure that they are giving a good description of their plan.

If you want to model real research, each group would submit its proposal to you and you would treat it as if you were a peer reviewer, that is, another scientist who would read the proposal and suggest changes. Peer review is done by all agencies that fund scientific research. You can make written comments and send it back. You can require a rewrite before "funding" their research.

Review the plans as a class, and let the students argue their experimental design. There are two ways you can do this. The students are likely to go for the simplest. Put clear water in one set of jars and sediment water in the others and put them in the same cool light condition. A much nicer design would be to shine the light into the jars inside two separate aquariums—one with clear water and one with suspended sediment. If the students come up with both designs, have them debate the relative merits of each. Why is one better than the other? Ask them to think about it. Are there variables other than just the dirt particles? What else comes from dirt? The jars with dirty water will have more nutrients and a different pH. What did they learn about the effects of acid on plants? Might the nutrients in the soil affect the plants? What if there were bacteria living in the soil? Might bacteria use oxygen? Sometimes there are blue green bacteria in the soil that might make oxygen. If the plants are placed in sealed jars of aged tap water, then all these variables are eliminated because the dirt is outside the jar.

Here are the two ways this experiment can be done by the class:

Outline

before class

1. mix soil and water sufficient to fill one small aquarium

2. boil and seal water sample

during class

1. lead student discussion of Activity 40

2. propose a different question: what happens to photosynthesis in plants in muddy water

3. how could they test this?

4. review variables and experimental design

5. design data tables

6. put jars with Elodea and aged tap water in the dark

7. remove the next day and test for dissolved oxygen

8. replace sample with boiled water

9. place jars in clean and cloudy water in aquariums

10. leave in light for at least 3–4 hours

11. remove jars and test for dissolved oxygen

12. compare results

13. write reports

	Plan 1	Plan 2
jars	dirt in aquarium water; *Elodea* in jars in aged tap water	dirt in water inside jars of *Elodea*
2 jars with plants	in aquarium with sediment water	sediment water in jar
2 jars with plants	in aquarium with clear water	clear water in jar

The problems with uncontrolled variables eliminates the second as good design. You might want to let groups try both and compare the relative merits of each design. Have students put their results on the board so that all the students can share the data. Have the students design their own data and reporting sheets and write up their findings as the publication of their research. If you can borrow or have scientific journals, you might want to let them see what the real thing looks like.

Results and reflection

There should have been more photosynthesis in the clear water. What if the sediment and the clear were both high—over 10–12 ppm DO? Then you may have left them in the light so long that both reached saturation, though the rates may have been different. How could this be tested? Repeat the experiment, but leave the jars in light for half as much time.

Have the students reflect on their experimental designs. Could they suggest changes that, if they were to repeat the experiment, would have improved the quality of the experiment? Compare their data charts and reporting techniques. What was the most effective in communicating their results and conclusions?

Conclusions

The amount of photosynthesis done by water plants is directly related to the available light. Where light is blocked by sediment, photosynthesis is reduced. Human actions that increase sediment in aquatic systems reduce plant growth and available food.

Extension and applications

1. Have your students list all the things they can think of that make water more turbid. Some things are sediment from farms or housing developments, sewage, and industrial wastes. They might not realize that things like plant nutrients that come from fertilizer, animal manure or sewage can make the water more turbid by encouraging phytoplankton to grow too fast. The phytoplankton can actually get so dense that they keep light from reaching aquatic plants growing on the bottom.

ACTIVITY 42 Competing for food

What is the relationship of food availability to the number of herbivores an area can support? Students model animal behavior in a simulation game.

Teacher's information

This game is a SIMULATION showing how food availability can limit the numbers of the animals that feed on that food. In this game or MODEL your students are the animals that are COMPETING for food. They will be zooplankton searching for phytoplankton in a small pond so they are HERBIVORES. The limits may be expressed in several ways. In extreme cases, the herbivores may starve to death. In the wild, animals that are suffering from lack of food frequently fall prey to predators or disease before they starve. Predators prefer weak animals that are less effort to catch. If there are no predators, however, food availability frequently limits the number of herbivores in a population.

Another result of insufficient food supply is that the number of offspring an animal has decreases. The result of the combination of increased MORTALITY (death by whatever cause) and decreased BIRTH RATE caused by low food supply is a decline in the number of animals COMPETING for the food. Lower competition means there will be more to go around for the next generation. The number of animals that an area can support on a permanent basis is called the CARRYING CAPACITY.

Introduction

Ask your students to recall Activity 29 in which they learned the words ZOOPLANKTON and PHYTOPLANKTON. If you did not do this lesson, review these words. Make it clear that many zooplankton feed on the tiny phytoplankton that drift through water. Explain that the students are going to pretend that they are zooplankton that eat phytoplankton in a pond and see what happens when herbivores have to compete for food. Have the students predict what might happen to them if they do not get enough to eat. Generally, they will predict starvation.

Science skills

organizing
inferring
predicting
experimenting
communicating

Concepts

Plant or phytoplankton (food) availability may limit how many herbivores live in an area.

When plants or phytoplankton are limited, herbivores compete with each other for their food.

Skills practiced

averaging
graphing

Time

1–2 class periods

Mode of instruction

teacher directed class work

Sample objectives

Students model a simple food chain.

Students explain how the availability of food can limit the population which depends on that food.

Builds on

Activity 29

Materials

for each student

10 markers—poker chips or plastic counters or other non-destructive small items; do not use seeds or items that break in half such as popcorn

plastic sandwich bag

worksheet

for the teacher

two data sheets

a clipboard

pencil

whistle or loud voice

Preparation

Plan the location for the activity, including an alternate site in case of rain if you are planning to do it outside. Do not try this in a classroom. Working around furniture is dangerous. This exercise works best in an open space such as the gym or outdoors. Count out the markers and sort in bags. Put 50 markers and 4 sandwich bags in a 5th bag and seal until you have enough for 10 chips per student for 3/4 of your class.

Action

Start with a pond that produces a limited number of phytoplankton and zooplankton. There will be the same food supply from one generation to the next. The phytoplankton are the counters or poker chips. Scatter the chips (10 per student playing the first generation), using whole bags of 50 chips. Count the plastic bags emptied and designate that number of students as zooplankton by giving each student one of the bags to collect phytoplankton. The rest of the class are the reserves and their chips will not be used. Designate the boundaries of the pond around the scattered chips. Have the students with bags stand outside the pond. You will always use this number of chips, regardless of how many offspring there are.

Tell the students that each will try to eat as much as he/she can without taking any away from another zooplankton. To slow the students down for safety, one foot must stay on the ground at all times; that is, each student must drag one foot, never letting it leave the ground. Say "go" and let them pick up all the food. It will be over fast. Use the whistle to stop the action. Have everyone sit down. Did each get the same amount of food? No, some individuals are more efficient searchers than others. Record the results on your data sheet and tell each group how many offspring they had.

Some of the zooplankton did not get enough food to reproduce while others did. Those that got more food left more offspring. Herbivores that got fewer than nine phytoplankton starved to death. Those that got nine to eleven left one offspring. With twelve or thirteen phytoplankton, they make two offspring. More than thirteen, they have three offspring.

After reproducing, the parents die, leaving behind the number of offspring indicated on the data sheet which are the zooplankton of the next generation. Change the number of players to match the number of offspring. Recruit from the reserves and/or allow substitutions for tired "zooplankton." Scatter the same number of food items that you did in the first game regardless of whether the number of players went up or down. Assume the same number of phytoplankton will be produced. The amount of food is limited. Run the game again and calculate the results.

Repeat. If possible, do four or five generations. You should find that as long as the food supply remains the same, about the same number of animals are produced in each generation.

Results and reflection

Back at their desks, have the students discuss what happened. Put the results of the simulation on the board. Ask them to give you some answers to the following questions. Did all the zooplankton get enough to eat? No; even though there were food items enough for everyone to survive and reproduce, some were better at competing for food than others. What happened to those that got more food? They used their extra food to reproduce. The best competitors had the most offspring. Did anyone find a particular trick to help him/her compete more successfully for food? If this trick or adaptation was one that was inherited, would their offspring also be better at finding food? Would more animals in the next generation be better at catching food? Yes, because those that got the most food left the most offspring. Have the students complete the worksheet and answer the questions.

Ask the students to discuss how the simulation or model is like the situation it represents and how it is different. For example, would there only be one

Outline
before class
1. count and package poker chips or other items

2. pick a safe site for game

3. duplicate worksheet

during class
1. have students predict effect of food (phytoplankton) availability on zooplankton numbers in a pond if the food were limited

2. scatter food

3. run the game with 3/4 of the class as zooplankton with 10 chips per individual

4. count chips collected and do data table as a class

5. repeat with numbers of offspring from the first game as adults in the second

6. fill out data sheet and repeat several more times

7. return to classroom and share data

8. do analysis and discuss

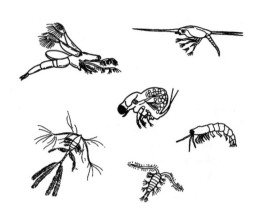

kind of zooplankton in a pond? Probably not. Did our pond have anything that ate zooplankton? No, but in a real pond there would be. Why is it useful to use a model? We developed a model in which food was the only thing that limited the number of animals so that we could study competition among animals of one kind and test different conditions under which competition took place. In a real pond we would have had too many variables to easily ask questions about population size in relation to food availability.

Conclusions

Animals compete for food. Those that do not get enough to eat do not reproduce or die or are caught by predators because they are weak or diseased. Those that compete most successfully leave more offspring. If the limit to the number of herbivores in an area is the food supply, their number remains more or less the same from one reproductive period to the next if the food supply remains constant. This average number of animals is the CARRYING CAPACITY for that habitat. If predators are added to this system, the number of herbivores may be reduced so that food is no longer the thing that limits the population. Under these circumstances fewer animals starve.

Using your classroom aquarium

Do you have any plants in your classroom aquarium? Any planteaters? If so, how well does the balance work between the two? Does one ever "win"? Sometimes algae in the tank may threaten to take over and look like the creature that ate New York. How do you control it? Add herbivores.

Extension and applications

1. Once you have your data, you might allow the students to change the rules, one rule at a time, and see what happens to the population. Ask the students to think of one change that would increase the number of zooplankton that survive. An immediate answer would be to increase the amount of food. Multiply the amount of food by 1.5 and run the game several times. Eventually a new herbivore limit is reached, although at a higher level.

Teacher's data sheet for Activity 42

Generation number _____
Total number of adults competing for food _____

no. adults getting this much food	number of food items captured	number of offspring	number of offspring by this group
	fewer than 9	0	
	9–11	1	
	12–13	2	
	more than 13	3	
Total offspring			

Generation number _____
Total number of adults competing for food (offspring from last game) _____

no. adults getting this much food	number of food items captured	number of offspring	number of offspring by this group
	fewer than 9	0	
	9–11	1	
	12–13	2	
	more than 13	3	
Total offspring			

Generation number _____
Total number of adults competing for food (offspring from last game) _____

no. adults getting this much food	number of food items captured	number of offspring	number of offspring by this group
	fewer than 9	0	
	9–11	1	
	12–13	2	
	more than 13	3	
Total offspring			

Activity 42
Competing for food

Name _____

1. Graph the number of zooplankton offspring you got each time you repeated the game.

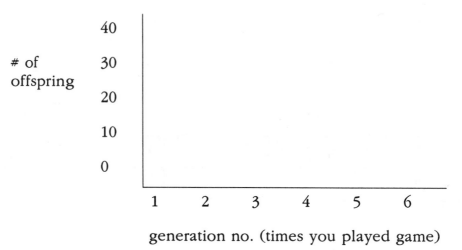

2. Calculate the average number of offspring in the population by adding the numbers from each generation together and then dividing by the number of generations you had.

3. The average number of offspring was _____.

4. Draw a dotted line across the graph to show the average. This is the average CARRYING CAPACITY for your pond. Describe how the actual numbers compare to the average.

5. Describe one change you could make in the game that would make the number of offspring increase.

6. Describe one change in your "pond" that might make the number of offspring decrease.

Activity 42
Competing for food

Name ___possible answers___

1. Graph the number of zooplankton offspring you got each time you repeated the game.

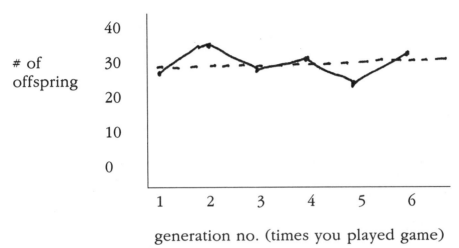

# of offspring

40
30
20
10
0

1   2   3   4   5   6

generation no. (times you played game)

2. Calculate the average number of offspring in the population by adding the numbers from each generation together and then dividing by the number of generations you had.

3. The average number of offspring was __27__.

4. Draw a dotted line across the graph to show the average. This is the average CARRYING CAPACITY for your pond. Describe how the actual numbers compare to the average.

The numbers of offspring in each generation moved up or down but stayed near the average or carrying capacity. They did not move very far.

5. Describe one change you could make in the game that would make the number of offspring increase.

We could have increased the amount of food (chips for phytoplankton) that we were eating. More food would have given us more babies.

6. Describe one change in your "pond" that might make the number of offspring decrease.

Less food would have caused the offspring to go down.

ACTIVITY 43 Eating and being eaten

What are some of the feeding relationships among the plants and animals that live in a pond? Students model animal behavior in a simulation game.

Science skills

organizing
inferring
predicting
experimenting
communicating

Concepts

The numbers of predators and of prey have a direct relationship which is a result of the way in which energy passes through the food chain.

Predators and prey have specific ways or strategies for dealing with each other.

A pyramid may be used to represent the feeding relationships in a food chain.

Time

2 class periods

Mode of instruction

teacher directed whole class work

Sample objectives

Students produce a model of a pond food chain.

Students develop feeding strategies appropriate to different levels of the food chain.

Students suggest changes in the rules of the simulation which test new ideas about feeding strategies.

Builds on

Activity 29
Activity 42

Teacher's information

This simulation builds on Activity 42 by adding the next levels of the food chain, those animals which feed on other animals: PREDATORS. The activity may be broken down and taught on several days, starting with the introduction to the concept of food chains or webs. (Most practicing ecologists use the term food chain to cover food web also.) There are a number of possible variations which the students may want to try once they have begun to play the game.

Introduction

Get settled either indoors or outdoors and explain that the class is going to pretend to be the animals in a pond. Ask the students to name some animals they might find in a pond. Big and little fish, frogs, crayfish, tiny zooplankton, insects, beavers and raccoons are some answers that are possible. Have pictures of as many pond animals and plants as possible and an enlarged food web diagram to illustrate these species. Explain that in our pond we are going to have groups of kinds of animals that represent levels based on who eats whom.

Where does the food come from? From plants which use light to do photosynthesis. Some of these plants are tiny phytoplankton while others are rooted green plants that grow under the water or along the edge of the pond. Write the words phytoplankton and green plants at the bottom of the board. Show pictures of pond plants and phytoplankton. Introduce the word PRODUCERS for those things that make food. Who might eat these plants? The tiny animals called zooplankton eat phytoplankton as the students learned in Activity 42. Many insects and the crayfish also feed on the plants. Write insects, crayfish, and zooplankton above the plants on the board and draw an arrow up to them. Explain you are drawing a diagram of the path food takes in the pond. Those animals that eat plants are called HER-BIVORES.

Who eats the herbivores? The little fish and frogs as well as some bigger fish. Add them to the next level along with an arrow. Animals that eat other animals are called CARNIVORES. The animals that get eaten are called PREY. Finally, who eats the carnivores? The big fish and the raccoon who are the TOP CARNIVORES. Do they just eat the level below them? No, they also eat the crayfish from the lower level. Can the students see the FOOD WEB or FOOD CHAIN forming as you draw the lines between the levels?

Action

Now for the game. The students are going to play a game which is really doing an experiment. Just like an experiment, there are rules to make it a fair test. They are going to be the animals in a pond food chain and are going to feed on each other. The phytoplankton (poker chips or plastic counters) are scattered over a wide area. Unlike Activity 42, in this game they are going to assume that there are more plants than the herbivores can eat: that food is not limiting.

Assign one third of the class to be zooplankton, crayfish and insects. Give them the one colored strip of cloth to wear and a plastic bag. They are all HERBIVORES. Scatter 20 food items for each herbivore. That means 20 times one third the number of students in your class or 200 items (20 x 10) if you have a class of 30. The chips are the phytoplankton in the pond. The herbivores must get their food by picking it up. To live, they must get 10 pieces of plant food before the end of the game. If they do not get 10 pieces before the game is done, they have died of starvation.

Give another third of the class a second color arm band or sash and a plastic bag. They are the CARNIVORES, the frogs and small fish. To eat, they must tag a herbivore. The herbivore gives up his food bag and sits down as he has been "eaten" by the predator and is out of the game. The predators must collect 20 pieces of food from the herbivore food bags to

Materials

for each student
20 markers (poker chips or plastic counters)
small plastic bag
for the teacher
plastic or crepe paper flagging or cloth strips in three colors
whistle or loud voice
pad or chalk board and pen or chalk
copies of the data sheet
food web diagram enlarged
optional
pictures of pond animals

Preparation

Do Activity 42 prior to this exercise. Pick a location to do the exercise with alternate plans for bad weather if you are going to be outside. See Activity 42 for discussion about safety. Collect the "flags" and cut them to either tie around a wrist or a waist.

Outline

before class

1. pick a safe location

2. enlarge food web diagram

3. collect and count chips and cut up flagging

during class

1. discuss the nature of the simulation and what the students will be modeling

2. start first run with 1/3 of class as each level of the food chain: herbivores, primary carnivores, top carnivores; use arm bands to designate roles

3. scatter 20 food items per herbivore each time

4. run first game: small fish tag zooplankton to get their food and big fish tag small fish to get their food

5. did anyone get enough food: herbivores need 10, small fish need 20 and big fish need 40

6. balance is not correct; have students change one thing and run again

7. repeat until some of each level of the food chain survive

8. return to class and share information; discuss results

be alive at the end of the game. They must stop eating when they have passed the 20 item mark. If they do not get 20 pieces before time is called, they have died of starvation.

The remaining students are the TOP CARNIVORES. They get the third color of cloth and a plastic bag. They feed by tagging either the herbivores or the carnivores who must give up their bags and sit down when tagged as they have been "eaten." The top carnivores need 40 pieces of food to be alive at the end of the game. They should stop eating when they have passed 40 food items.

Caution the students about rowdy behavior and running into each other. Make them drag one foot on the ground at all times to slow them down. Give them several minutes to play the game following the above rules and then stop them.

Results and reflection

Sit down and analyze what has happened. Use the board and data sheet. Do the students think this model worked like a real pond food chain? What was the cause of death in most cases for the herbivores? For the carnivores? Were the proportions the same at the end as the beginning? What levels of the food web do they think should have the most animals?

The original proportions are intentionally wrong. They were chosen because they do not work. The students should be able to see the lack of balance. The top carnivores are going to eat all their prey. With no prey left to reproduce, the top carnivores will starve to death. The numbers at each level should be more like a pyramid. If all the herbivores get eaten, they will not leave any offspring, and the rest of the levels will starve to death in the future.

Action again

Let your students discuss these questions. Then let them pick one thing to change about the original rules, other than changing the amount of phytoplankton. They can only increase the phytoplankton by increasing the number of zooplankton. Continue to scatter 20 pieces for each herbivore. If they decide to have twice as many herbivores, then they can also have twice as many chips. Run the game (the experiment) again from the beginning. Would it be a "fair test" if they changed more than one thing at a time? No. In order to compare the first with the second experiment there can be only one difference (variable). Otherwise they cannot tell which change caused the different result. Here are some changes that they might make:

1. Change the number of herbivores or top carnivores. If the proportions are changed by deleting some students, you can keep within the rules of changing only one thing at a time. If you delete some in one run, they can be added in the next run. For example, most of the top carnivores can be removed. After seeing what the result is, they might be added as herbivores in a subsequent test.

2. Give the herbivores some places where they can hide or are safe from the carnivores. Draw a safe zone with string or rope or use something like a hula hoop. They can run out to grab a bite to eat and then hide. But remember, they have to get 10 pieces of food or they die of starvation. An example might be the burrows in which crayfish hide.

3. The primary carnivores (small fish) may also be given a refuge in which they are safe from the top carnivores (big fish). Have the students give you some guesses as to how frogs or small fish could hide from predators. Both might be able to hide in vegetation. The food might be in the open while tagging a tree or pole is "safe." Another possibility which protects the herbivores is that some feed when the predators are not active such as at night. Have 30 second safe times periodically when the top carnivores rest and stop eating. In actual fact, many predators catch their prey at dawn and dusk.

Repeat one change per game until the class has created a balanced pond for the food chain. In order for the food web or chain to be realistic, the students must have some individuals from each level alive at the end of the model. These animals will be the ones that reproduce, making the next generation. Make adjustments until this happens. It requires very few top carnivores (perhaps one), a few carnivores and lots of herbivores.

Results and reflection again

What proportion of herbivores to carnivores to top carnivores worked? If you drew the numbers of students in each group in the most successful game as blocks with one block equal to one animal and made three rows with the herbivores on the bottom, the carnivores in the middle and the top carnivores(s) stacked on top, what would it look like? Draw it. Does it resemble a pyramid? It should. A food chain can be represented by a pyramid with the bottom as the lowest level of the food chain.

Have the students discuss the strategies they used to catch prey while not being caught. The smart top carnivore will let his prey alone until they have eaten enough to be of value when caught. Have them consider how accurate and useful they think this model of a food chain was in demonstrating the concepts. Can they continue to suggest changes and improvements in the model? Scientists create a model. Then they test the model by asking questions with it in the lab and by comparing their lab answers to studies of the real world. They work until their model seems to function like the real world. In doing so, they have to examine the parts and their relationship to each other in detail. For example, there are several big computer models of world climate. Scientists ask the models to predict future global weather such as El Niño events. Then they watch the real weather to see if the predictions were correct. Each time they "tinker" with the model to improve it.

Conclusions

This exercise illustrates that there are fewer animals in each succeeding level of the food chain. It also introduces the concepts of predator-prey interactions and the strategies that predators and prey use in feeding and hiding from predators.

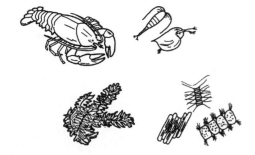

Using your classroom aquarium

Have your students identify the feeding level of each of the different kinds of animals in your classroom aquarium. Is your aquarium a balanced ecosystem with regard to predator-prey relationships? If you have more predators than prey, what do you do to enable the predators to survive? You feed them food produced outside the tank. What changes would have to be made to create a balance in your aquarium?

Extension and applications

1. Have the students write several paragraphs about how it made them feel to be "eaten" or to not get enough to eat and to "starve." Most wild animals must face these problems routinely.

2. If your class has computer software that includes food chain simulations, set one up so that the students may experiment with it. Have them compare their model used in class with that of the computer. Identify which animals were producers, consumers, etc. Did the computer model give similar results?

3. Humans have produced their own food chains. Have students do an analysis of common domestic farm animals with regard to their place in the food chain. It becomes apparent that most of the major animals we depend on for meat, milk or eggs feed low on the food chain. Why? Each animal uses food to produce heat and movement as well as for growth. Energy spent on heat, movement, digestion and chemical work inside its body is lost. Only energy which is used for growth is available to the next level of the food chain. Less food would be available if we fed on carnivores rather than herbivores. Examples: cows, sheep, horses, rabbits, geese, and goats eat plant material; chickens and pigs are fed primarily plant material, although they may also get fish meal or other animal products. The two carnivores that are common on farms are kept for behaviors that relate to their feeding habits. Cats kill rodents that would steal grain from the farmer. Dogs hunt by chasing things, making them naturals at herding cows or sheep. They also bark at the sign of intruders.

POND FOOD WEB

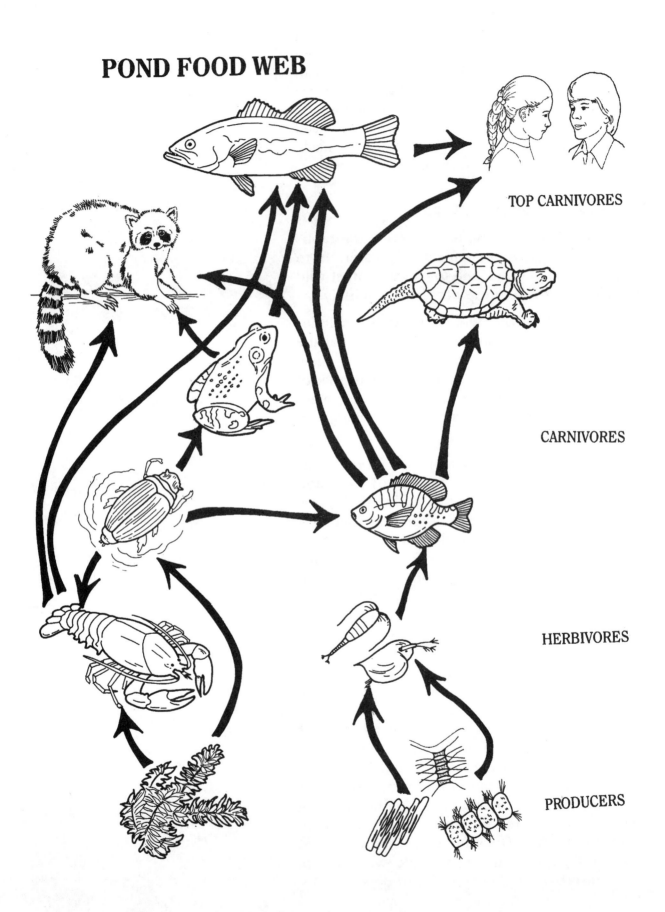

TOP CARNIVORES

CARNIVORES

HERBIVORES

PRODUCERS

Teacher's Data Sheet for Activities 43 and 44

First Run

feeding level	# alive at start	# eaten	# that starved	# that survived
herbivores				
carnivores				
top carnivores				

Second Run

feeding level	# alive at start	# eaten	# that starved	# that survived
herbivores				
carnivores				
top carnivores				

Third Run

feeding level	# alive at start	# eaten	# that starved	# that survived
herbivores				
carnivores				
top carnivores				

Fourth Run

feeding level	# alive at start	# eaten	# that starved	# that survived
herbivores				
carnivores				
top carnivores				

Fifth Run

feeding level	# alive at start	# eaten	# that starved	# that survived
herbivores				
carnivores				
top carnivores				

ACTIVITY 44 Getting caught

Do some human activities change the feeding relationships, and thus the ecological balance, of a food web or chain? Students model fisheries management in a simulation game.

Science skills

organizing
inferring
predicting
experimenting

Concepts

Removing the top carnivores from a habitat can change the feeding relationships at other levels of the food chain.

Human management of natural areas requires an understanding of the importance of a balanced food web.

Time

1 class period

Mode of instruction

teacher directed whole class work

Sample objectives

Students identify human activities which affect feeding relationships in an aquatic environment.

Students identify correct fisheries management practices.

Builds on

Activity 42
Activity 43

Teacher's information

In this simulation activity your students will learn how the basic ecological principles relating to food chains, which they have experimented with in Activities 42 and 43, can be applied to the management of aquatic habitats. Human activities may affect food chains in many ways. Pollution and habitat destruction cause obvious changes that are easy to see. Dead fish are hard to ignore. It is also easy to understand the consequences of filling a pond or marsh in order to build a shopping center or to add another field to a farm. The whole habitat goes away. Much more subtle changes may result from the extensive harvesting or removal of selective levels of the food chain or the addition of new species which change the relationships among levels.

When we selectively remove some fish, we are making changes that may have a significant impact on the other animals and on the plants in the pond, lake or ocean. Very careful planning of fishery regulations are necessary to be able to remove animals from a system on a long term basis without destroying the ECOLOGICAL BALANCE of the entire system. The result of this careful planning is a calculation of how many fish can be taken year after year without upsetting this balance, the SUSTAINED YIELD. The people who study these problems are in a field called FISHERIES or WILDLIFE MANAGEMENT. They set fishing limits for both sport fishing and commercial fishing and enforce them. They make use of practical applications of the kinds of ecological principles learned in this exercise.

Introduction

Explain that the class is going to continue to experiment with food chains, but this time the students are going to ask questions about the impact that humans can have on the balance of a food chain. Ask for some ways that some people might affect the animals in the pond food chain that they have been using as a model. While the students are most likely to think of things relating to pollution, someone will probably come up with fishing. Ask them to tell you what the feeding levels were in the pond model and what kinds of animals lived at each. Draw the food web on the board as they give it to you.

Which level is the most likely to be taken by humans? Since people like to catch big fish rather than tiny minnows, big fish are the likely candidates for human removal. The children are less likely to think of hunting for sport or trapping for fur. Another top carnivore, the raccoon, is frequently hunted with dogs as a sport. During the winter, raccoons are trapped for their fur which is used in fur coats. Ask if anyone has seen a coat made of or trimmed with raccoon fur. (It has long hairs and is a mix of tans or browns.) Therefore, in our model pond, we are going to assume that people come to the pond and catch all the big fish while other folks hunt all the local raccoons. We now have decided how humans are going to disturb the pond food chain in our model.

Action

Run the experiment. The first time around see what happens before humans are involved. People will be added to the system in a later run. Start by designating 3/4 of the students as herbivores—the zooplankton, insects, and crayfish. Make 2 students top carnivores—the big fish and raccoons. The remaining students will be carnivores that feed on the herbivores—the frogs and small fish. Hand out the colored flags which show their feeding level and the plastic bags to hold the food they eat.

Scatter 20 food items per herbivore. Each herbivore needs 10 to survive and reproduce. Each carnivore needs 20 pieces and each top carnivore needs 40 to survive and reproduce. When any animal has enough food, it can go to hide in a safe place. When any animal has been tagged by its predator, it has been "eaten" and must give up its food and sit down.

Allow the students to play the game until everyone has either gotten enough to eat, run out of food to eat or been eaten. Was this in reasonable balance?

Materials
for each student
20 markers (poker chips or plastic counters)
small plastic bag
for the teacher
plastic or crepe paper flagging or cloth strips in three colors
whistle or loud voice
pad or chalk board and pen or chalk
data sheets from Activity 43

Preparation
Collect the materials and select a safe site. See Activity 42 for a discussion of safety.

Outline

before class

1. gather materials and data sheets

2. pick safe location

during class

1. ask what will happen when the top carnivores are removed from the pond

2. start with 3/4 herbivores, 2 top carnivores and the rest primary carnivores and 10 food items

3. run the game and check the results

4. remove both top carnivores and use 3/4 herbivores and 1/4 primary carnivores

5. compare the outcome

6. discuss the data

7. discuss the role of models in decision making

That is, did some members of each level survive? Record the results.

Now repeat the experiment with one big change. A trapper came and caught the raccoon and a fisherman or woman got the big fish. Have the same children take each role, but the two top carnivores sit out. Have the students predict the outcome. Run the game again.

Results and reflection

What happened to the balance? Did the herbivores get eaten in larger numbers? What would the children predict would happen in the next generation? Ask the students if they can see the direct consequence of removing the top carnivores on the structure of the food chain. Even in one generation, there should be the obvious result that there is no longer a balance. The secondary consumers are now able to eat almost all the herbivores, leaving few to reproduce in the following year. With the herbivores gone, the carnivores will starve in the following years.

Ask the students if they think working with models of how a system works might help fisheries managers decide on fishing regulations? How could a model be used in the real world? If managers knew enough about food availability, numbers of different species, catch by fishermen, etc., could they test different fishing levels in their models and set the catch so that the fishing could be sustained through time? The really big models are done on computers, allowing very complex models to be developed. Students need to understand the value of models in decision making. We are becoming able to actually predict the consequences of various human actions. Political decisions need to be made on the basis of these models.

Conclusions

Natural food chains are generally balanced over a period of years. When all of one level is removed, the other levels will be affected. Any human harvesting of natural populations must take into account the effect on the other levels of the food chain. Top carnivores are an important part of the balance of the food chain, not bad things that should be killed.

Extension and applications

1. Have the students list top carnivores that are hunted in the ocean: tuna, billfish and large sharks are all being harvested much faster than they can reproduce. Can they predict possible outcomes? They can research the current problems with these large predators and over harvesting by contacting marine conservation organizations and the National Marine Fisheries division of the National Oceanic and Atmospheric Administration (NOAA) using the Internet.

2. For a comparison of what has happened in cases where people have actually selectively removed whole levels of a food chain, consider the following:

   a. In the early part of this century, extensive hunting of mountain lions in northern Arizona effectively removed predation from the deer herds. In the years that followed, the deer herds grew in number because their predators were gone. Eventually there were so many deer that they ate all the vegetation. Then the deer started starving during the winters. They had virtually destroyed the vegetation on which they depended. During succeeding years the deer population declined dramatically due to starvation. In this case removing the top carnivore resulted in starvation of its prey and such extensive damage to the plants that neither the vegetation nor the deer have ever recovered properly.

   b. Recently extensive fishing for small fish called capelin has so reduced the numbers of these animals that the sea birds such as puffins that depend upon capelin for food are experiencing reproductive failure as their young starve. It is possible to overfish capelin because they swim in tight schools. By hunting for the schools with planes and surrounding them with large nets from boats, humans can catch so many that there are not even enough fish left to reproduce.

puffins

Teacher's Data Sheet for Activities 43 and 44

First Run

feeding level	# alive at start	# eaten	# that starved	# that survived
herbivores				
carnivores				
top carnivores				

Second Run

feeding level	# alive at start	# eaten	# that starved	# that survived
herbivores				
carnivores				
top carnivores				

Third Run

feeding level	# alive at start	# eaten	# that starved	# that survived
herbivores				
carnivores				
top carnivores				

Fourth Run

feeding level	# alive at start	# eaten	# that starved	# that survived
herbivores				
carnivores				
top carnivores				

Fifth Run

feeding level	# alive at start	# eaten	# that starved	# that survived
herbivores				
carnivores				
top carnivores				

SECTION VI Wrapping it up: projects and programs

This section has activities that are long term projects or summary events that cap off the classroom use of *Living in Water*. Some of these are to be done independently as out of school projects based on library and Internet searches as well as other kinds of information. One is a research project, and one an engineering problem. Some are whole class events. All are assessments of student understanding of this curriculum. Be sure to save time for these.

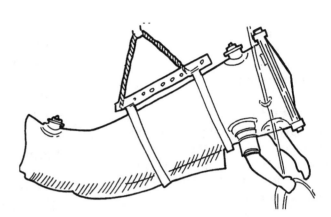

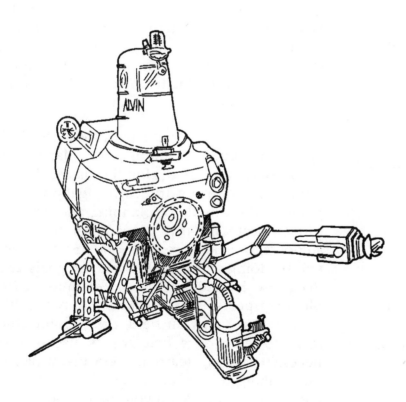

ACTIVITY 45 Getting wet!

An aquatic field investigation applying concepts and skills learned from this curriculum to the study of a real habitat.

Science skills

observing
measuring
classifying
organizing
inferring
communicating

Concepts

Ecologists study the physical and biological characteristics of an environment to try to explain the distribution and abundance of organisms in it.

Time

4–10 class periods

Mode of instruction

teacher and parent directed small group work

Skills practiced

graphing
averaging
testing water
collecting and quantifying organisms
identifying organisms
mapping and describing habitats

Sample objectives

Students plan and execute an ecological study of an aquatic environment.

Students collect and analyze data.

Students give oral and written presentations based on their field investigation.

Teacher's information

Nothing is more work than taking students on a field trip. Nothing is more rewarding than going out to study the real world. It is impossible to write a single guide for doing a field investigation. You may organize a three day extravaganza to the wild or spend a half a day working on a pond in a local park. If the site is on your school grounds, the study might cover several months of weekly measurements. Success depends on planning, planning and more planning.

The nature of the experience

There are two common problems with field trips. The first is that the people running the trip treat it as a mobile lecture. That is they talk, talk, talk, and the students never actually do any scientific work. Field trips should be largely student work, not leader lecture. Students should be making observations and measurements, collecting data, and identifying organisms, not listening to an adult tell them what stuff is and how the Indians ate it.

The second problem with many field trips is that they do not include classroom work planning the trip ahead of time or analyzing the data collected upon return: they are not true investigations. If you are going to the trouble of planning and organizing the trip, allocate pre-trip preparation time with your students to make sure everyone knows what he/she will be doing and why. Set specific goals and objectives. Involve students in the planning. Also, leave plenty of post-trip class time to study the things you collect or see on your field trip. Helping students to learn to learn on a field trip is also helping them become life-long learners. Too many people grow up thinking they need a person to tell them about the natural world. With your help, your students can learn that they have the ability to teach themselves about the natural world and to enjoy it all their lives.

Selecting a site

There may be a nice pond or stream within walking distance of your school which would make a great site. Ask your students. If there is a body of water nearby, some of your children will have found it. Streams are common, even in cities. Cemeteries and golf courses often have ponds. If you can afford to take a trip on a bus, you may go farther afield. If it is within your budget, consider using a nature center, environmental education site, marine lab or other organization that runs field trips. Working with professional field trip people will save you a good deal of time. They provide facilities and equipment. Call your state environmental science specialist, county science specialists or high school environmental education teachers for suggestions about places to go. City, county, state or federal parks, reserves or wildlife refuges may also have environmental programs.

Visit the site yourself first

Regardless of who is running the trip, visit the site yourself first. If there are staff members at the site, talk to them. Take a teacher workshop or training program there. Take slides to share with your students, make a map, and collect information on the habitat you will visit. Purchase field guides in the gift shop.

Questions to address

- Is it legal for you to use the site? How do you get permission?
- Is the site safe? What kinds of precautions should be taken?
- Are there rest rooms?
- What happens if it rains?
- Where can you eat lunch?
- Do you or the organization you are working with have adequate equipment? If not, can you borrow it from a local high school or other source?
- Do you have field guides for identification or can you check them out at the library?
- Have you made reservations early? Plan a spring field trip early in the fall.
- Have you arranged for transportation?
- Have you sent home permission slips for the trip?

- Have you recruited and trained your chaperones? In addition to parents, are there older students who can mentor yours on the trip?
- Have you given the students orientation to the site? Do they have a list of what to bring and what to wear? Are they divided into working groups, ready to go?

Prepare students ahead of time

Have the students do library research on the habitat and the organisms they can expect to encounter. Collect field guides that will help you identify what you find. If you are running the trip yourself, make sure you have permission to use the site and permission to collect if you plan to take fish or plants. Collecting permits may be obtained from your state departments of natural resources or fish and game. Show the permit to your students ahead of time and explain that one cannot just go out and collect without it. Actively involve your students in planning the field work. Give them maps of the site, examples of the equipment available, and bottles, jars and plastic bags to label for specimens. Have them help organize the gear, make the data sheets and plan the day. If you are going to do water testing, bring a bottle to collect used chemicals. Practice the tests ahead of time if you plan to do new ones, such as salinity, that are not in *Living in Water*.

Planning a project

What kinds of information do you want to collect on your trip? What activities will your students do? Consider trying to correlate physical characteristics of the environment with the distribution of species. Here are some projects you can do:

In a pond, lake or estuary:

1. If you have one or two seines (small nets that can be pulled between two people), compare the number of species and the abundance within species of fish swimming over sandy versus muddy bottom. Or compare the fish collected from an open site with those from a weedbed. If the site is homogenous, just do a fish census. Collect one of each kind of fish in a numbered plastic bag. Count all of the rest you find that

National Aquarium in Baltimore

match it and record their numbers. Back in the classroom, have the students identify the fish, tally up the total number of each species and graph them. Bar graphs often show different species numbers over sand versus mud or in the open versus cover.

2. If you have water test kits and a sampler, look for variations in the distribution of dissolved oxygen in places: test surface versus bottom water in a lake or estuary or day versus very late night samples in a small pond. A dock makes sampling easier. Also check the temperature top and bottom. If you are in an estuary, check salinity as well top and bottom.

3. With a plankton net, you can look for zooplankton. You do not need to have a boat. Walk along a dock or throw the net out and pull it back to you. Make sure everything is securely attached before you throw it though. Compare zooplankton at the surface and the bottom by weighting the net with fishing weights to get a deeper tow. Or just do surface tows if you can start at dawn. Even in a shallow pond, the zooplankton do vertical migration on a daily basis. Up at night and down during the day to hide from their fish predators. You can make a plankton net from cheap pantyhose, a coat hanger and two spice jars (see Recipes).

In a non-tidal stream
Do a biological survey of the insects and other invertebrates living on the stream bed. All you need is some plastic dish pans and guides to identification. The Save Our Streams material from the Isaak Walton League has good information on these organisms (called macrobenthic aquatic invertebrates). State or county departments of the environment may also have good guides or programs. Since these animals are good indicators of water quality, you can discover whether your local stream is healthy or not without doing chemical testing.

In a stream, pond, lake, river or estuary
If you have nitrate test kits, check for nutrient loading and have the students look for possible sources if there are nitrates in the water.

In a stream, pond or lake

If you have a pH test kit, check for acid water. A pH below 6.5 can affect aquatic life. A pH below 6 is serious. Check for aquatic organisms. Compare with a neutral water site. Search for sources such as acid mine drainage or acid rain. Do not bother with pH in salty water. It has a natural buffer.

In a wetland or at the shore

Transects across ecological gradients reveal changes in species composition. What happens to plant species as one moves from dry land into a marsh or pond? What changes are seen in invertebrates and seaweeds from the very high to the low intertidal zone? Use a marked rope to record distance. Record changes in species and count how many of each there are. Be careful not to destroy what you are studying.

Be environmentally responsible

Never collect without permission of the landowner if on private land or the authorities if on public land. Collection without a scientific permit is forbidden in many areas. Always find out what is legal ahead of time and explain the laws to your students. Collect as few plants and animals as possible. Do not dump test chemicals into natural bodies of water. Teach your children respect for each organism. Do not have each student do a large collection of whole plants or animals. If plants and animals are needed for later classroom identification, take one of each and leave all the rest where you found them. Label each collection completely so that it can be used in class. Always take heavy gloves and trash bags. Have students clean up the site before you leave.

Be safe

Handle chemicals yourself if you have younger students. Supervise their use by older students. Practice water safety. Everyone wears a life jacket when on a boat or even on a dock. Take sun screen. Know what poison ivy or poison oak looks like. Know where a phone is and take emergency numbers with you. For students with medical problems, know what they are and encourage their parents to act as chaperones. They do not have to work with their own children's group, but just be there if needed.

Back at school

When you return to the classroom, identify what you found. There are field guides for almost everything. For example, for studies of ponds, the Golden Guide *Pond Life* is unbeatable. Look at small things under a dissecting microscope. Analyze the data you collected. Make charts and tables to display data. Average them. Graph them with line or bar graphs. Look for correlations. Have working groups write reports and give class presentations on their results. Have oral presentations for the parents' science symposium or give presentations to other classes.

Take it a step further

Did you discover an environmental problem? Discuss with your students what you might do to improve the field site you visited. Set them up to do research on who might help and how. Work with an environmental professional and your students to plan and do an environmental project to improve the site.

Extension and applications

1. Have a field biologist talk to the class about what kinds of work he/she does. Thanks to television, students have very strange ideas about field studies. The rain, heat, cold, mud, insects, long hours and tedious data analysis are missing on nature programs. Look for people from your state park system, state department of natural resources, a local environmental consulting firm or nearby college. Avoid high profile institutions like zoos or aquariums where they get dozens of such requests a week. Seek people actually conducting field research. Perhaps you could ask this person to help your students plan the field trip or an on-going research project.

ACTIVITY 46 In the life science research lab
Research projects that are designed and conducted by students.

Science skills
measuring
organizing
inferring
predicting
experimenting
communicating

Concepts
Scientists plan and conduct research based on previous research outcomes.

Skills practiced
measuring dissolved oxygen

Time
5 or more class periods

Mode of instruction
independent individual or group work

Sample objectives
Students design and conduct an independent research project.

Builds on
Activity 15
Activity 17
Activity 21
Activity 22
Activity 40
Activity 41

Teacher's information

In the experiments done throughout most of *Living in Water* the questions addressed have been posed by the authors and the experimental design has been given to the students. By now, they should be thoroughly familiar with controlling variables, multiple repeats of tests, data collection and analysis. They should also be very good at testing dissolved oxygen. This is their chance to design their own experiment. Rather than turning them loose in the standard science fair style where anything goes, this exercise gives them equipment and organisms with which they are familiar and the chance to ask questions related to work they have already done. This models the experience of real scientists much more closely than the totally open-ended science fair project typically copied out of a library book with parental help. These experiments can grow into a full blown science fair project, but they start on familiar ground.

Introduction

With the equipment on display, review the kinds of biological questions the students previously asked with it. What can they measure in the aquatic plant *Elodea* with these materials? Respiration and photosynthesis. What are some of the kinds of questions students could ask using the materials you have provided? Lead a brief discussion of the possible ideas. Now how would a scientist proceed?

- Ask a question.
- Do a library literature search to make sure it has not already been answered.
- Predict what she/he thinks the answer might be.
- Describe why this might be an important question to answer.
- Design an experiment to test the question.
- Do repeated tests and collect the data.
- Analyze the data and compare to the prediction.
- State the conclusions.

Action

Each student, pair of students or group of students, depending on how you organized the class, will do the tasks listed above, with the probable exception of the literature review. Have the students write out their research plan and trade them with other groups for review. This is much like the process of submitting a grant proposal to do research. The proposals are graded by peers who also make suggestions for improvement. Then submit rewritten plans to you for review before beginning any work. You may request that it be re-written, based on your written comments and questions. The students may refer to your copy of *Living in Water* to review methods for measuring photosynthesis and respiration.

Here are some of the kinds of questions that can be asked with the equipment and plants listed above:

- What is the effect of a range of several concentrations of a chemical (i.e., table salt, dish washing detergent, plant nutrients) on respiration or on photosynthesis in *Elodea*?
- What is the effect of different temperatures on photosynthesis (use water baths of different temperatures in one light regime)?
- What amount of light (what distance from a light source) is sufficient to produce maximum photosynthesis?
- What is the effect of keeping plants at different temperatures over a week or two on the rates of respiration or photosynthesis; do they acclimate?

Results and reflection

Have the students write up their work. For models, let them see some real scientific papers if you have them. Their "papers" should include the following sections:

- abstract: a very brief summary
- introduction: the question asked and their prediction
- methods: the methods which they used to test their prediction in enough detail that a reader can repeat the work
- results: the data and some method of data display and analysis
- discussion and conclusions

Materials

for the class

dissolved oxygen test kits
Elodea
pint and/or half pint glass canning jars
aged tap water
graduated cylinders
measuring spoons
balances or scales
vinegar
LaMotte Nutritabs (see Recipes)
a variety of safe household chemicals that might end up in water such as salt, detergent, shampoo, etc.
grow lights or indirect sunlight
clear plastic aquariums large enough to hold the canning jars for temperature control
ice
thermometers

for each student
goggles
optional
copies of real scientific publications

Preparation

Assemble all of the equipment where the students will be able to see and handle it.

Outline

before class

1. collect the materials

2. collect scientific papers

during class

1. have students review past experiments using the equipment listed

2. review the process scientists use in doing an experiment

3. have students work on designing and writing a plan for a new experiment using this equipment

4. use teacher and peer review of plans

5. refine plans and do experiment

6. work up data

7. write paper and abstract with peer review

8. reflect on writing

9. rewrite

10. do poster and publish papers

Work with the class before they start writing to design a list of the characteristics of good writing to communicate information. There is a common misconception that scientific writing is dull, turgid and written in third person. In fact first person voice is entirely acceptable as in "We did the following tests . . ." rather than "The following tests were done . . ." Perhaps the hardest task in the list above is writing the abstract and yet it may be the most important for a scientist. Others read the abstract to decide if the paper is worth reading or just read the abstract and skip the paper if it is in a field that is not directly related to their research. If you have scientific papers for the students to look at, you might have them examine them and then try to tell you what the purpose of the abstract is. Then have them practice writing an abstract of an experiment they have already done for *Living in Water*. Let them use the peer review process described below. Then ask them to reflect on how they approached the task of writing an abstract. Have them explain what process they used in thinking about writing it. Let them compare writing an abstract with writing a summary of a movie, a story or a television program. The abstract of their work will be published in the printed program if you choose to have them present talks or posters about their research for other students or for their parents.

Have them give their papers to other students to review and comment on. This is also called peer review. All scientists' work is peer reviewed before publication. Scientists are not paid to help other scientists by doing peer review. It is a part of their professional obligation. Allow the students to rewrite their work or even repeat it in a modified way following peer review.

Have the students discuss and reflect on the value of having to write their plan of action down before doing it. What was the value of having them reviewed by others? Did they clarify their thinking and refine their plan? Did it save them time?

In addition to their paper, have the students generate a poster about 2' x 3' which they can use to tell their peers about their work. It should have both text and figures with tables, graphs etc. Or they may prepare a 5 minute talk on their work. Both talks and poster sessions in which students stand by their posters and discuss their work with others are popular ways of communicating among scientists.

Conclusions

Scientists spend more time planning, thinking, working up data, analyzing data, and writing than they do doing experiments.

Extension and application

1. Encourage successful projects to be expanded and developed as science fair submissions.

2. Have the students present their work for other classes or for their parents.

3. Invite a scientist to come to your class to discuss her/his work and bring examples of experiments, data, publications etc.

4. Consider publishing a class scientific journal for distribution to other classes or parents. Or put students' papers on the Internet for other schools to read. Electronic scientific publication is a very rapidly growing concept. Printing costs are high, and publishing in print is slow in a "hot" field. Submit some of the papers to publications like your state science teachers association newsletter or have students write up their work as experiments for other students to use. Send their work to science teachers journals like *Science Scope* from the National Science Teachers Association.

ACTIVITY 47 Sight-sea-ing!

Designing a tourist ride to the deep sea. Students apply knowledge about the physical characteristics of water and marine environments gained during class to an engineering problem.

Science skills

organizing
inferring

Concepts

Assessments may be in the form of application of knowledge to solving a new problem.

Engineers must take all physical characteristics of the environment into account when designing new things.

The marine environment has complicated sets of physical characteristics which challenge human occupation of the oceans.

Skills practiced

drawing

Time

2 class periods or homework

Mode of instruction

individual student work

Sample objective

Students design a tourist submersible ride that correctly applies an understanding of the physical characteristics of the ocean to an engineering problem.

Builds on

Activity 25
Activity 30
Activity 31
Activity 34
Activity 35
Activity 36
Activity 37

Teacher's information

By the end of this curriculum, your students should have a fair understanding of what the physical characteristics of the deep ocean are: it is cold, dark and has high pressure. Additionally, they should have an understanding of the characteristics of water and have begun to address buoyancy and the difficulty of maintaining position or moving through water. A thorough understanding of buoyancy may well be developmentally beyond early middle school students. This exercise asks students to apply their knowledge to an engineering problem—designing a tourist submersible ride. It also asks them to think about the basic needs of humans and apply this understanding to their designs as well.

Introduction

You may choose to have a class discussion reviewing what the students have learned during *Living in Water* or you may give this assignment "cold." You may want to develop a scoring rubric with your students as a way of discussing what constitutes a complete project. Avoid the word submarine when discussing this activity with the students as the word suggests specific designs.

Action

Give the students the worksheets and have them work silently. If assigned as homework, there may be much more elaborate projects, but there may also be parent intervention. If this is an assessment task, class work is more fair.

Results and reflection

When finished, have the students list the physical characteristics of the ocean that they had to keep in mind. What were the design constraints that had to do with the economics of the operation? What did they have to think about with regard to human needs? Post the students' projects so that they can be shared. Here are some things they should have considered.

Physical things about the deep sea:

- light: it is dark underwater so you need lights, but they need to be turned off sometimes for tourists to enjoy the bioluminescent animals
- pressure: the submersible needs to have strong walls
- moving through water: the design may be somewhat streamlined, but it does not have to go fast so a sphere or flattened sphere is fine
- buoyancy and density: close to neutrally buoyant; the cheapest way to go up and down is to use compressed air to change the buoyancy, just as a swim bladder does in a fish; fill air spaces with water to go down and displace the water with compressed air to go up (needs to be very compressed) or to go up, turn on a small motor
- cold: it is cold in the deep sea as cold water sinks
- staying in place or moving about: some will design a ride that goes up and down on a cable and some will make a "submarine" that moves around
- hot water: a deep vent spews forth very hot water

Human needs:

- an air supply
- a good view
- bathrooms
- drinking water
- seating
- safety

Economically useful:

- food service
- a gift shop
- fuel economy

Conclusion

There should have been an interesting range of solutions to this design problem.

Materials
for each student
worksheet
drawing materials

Preparation
Copy the worksheet.

Outline
before class
1. copy the worksheet

during class
1. give students the worksheets and have them work independently

2. lead discussion of physical characteristics of the ocean they had to consider

3. what human needs did they have to address

4. discuss economic considerations

5. share designs

Extension and applications

1. Research the real tourist submersibles that are being built. Compare their design to the students'. In what ways are they the same? Different? The students may well have a better design.

2. What about real submersibles used in scientific research? Use the library or the Internet to research both "manned" submersibles like the *Alvin* and remotely operate vehicles (ROVs) like *Jason*.

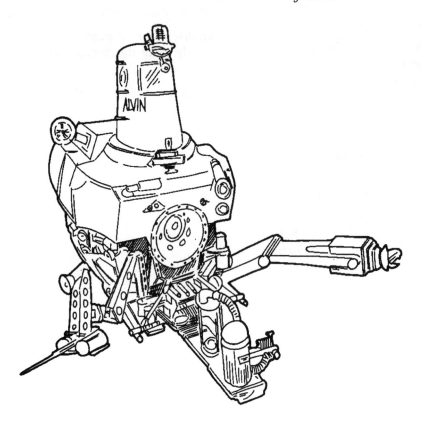

Activity 47
Sight-sea-ing!

Name _____

Working alone

You are planning to open a radically new kind of tourist attraction—a slow ride into the deep sea to view the wonderful bioluminescent fish as well as weird invertebrates which only a few scientists have seen in person! At the bottom of your site is a deep vent community of giant tubeworms which you also want people to be able to see.

You want it to be very safe and comfortable, but you also want to make a lot of money by designing something that is cheap to run. You want it to hold a reasonable number of people (40) with each having a good view of the underwater life forms you might encounter. You will make two trips under water each day, going down to 2,000 m. You want it to sink fast through the surface water and then very slowly through the deeper water. You do not want to spend money on a lot of fuel. When you get down to 2,000 m near the deep vent, the submersible should remain there for a while and then return to the surface reasonably fast. You have a deep water location in an underwater canyon near shore. You must also plan for how people will get into the submersible from shore.

You want to have all the things humans need to survive underwater as well as the special things that people on vacation expect to make them happy. You also want to have a profitable operation that makes you and your company a good living.

1. List the things you must consider about the physical characteristics of water and the ocean in order to make a successful design. Think about all the things you have learned about water and about moving through it as well as what it is like in the deep sea.

2. List the things you will need to provide for the people who are going to ride in your submersible.

3. Explain how you will make the ride go up and down and stay in one place.

4. On the next page make a detailed drawing of your tourist attraction. **Label the parts** that have important functions related to the lists above. Consider a cross sectional drawing. Any additional things you want to point out will be welcome.

5. Describe how you applied two things you have learned about the deep sea to the design of your submersible tourist ride.

Drawing of your tourist ride.

ACTIVITY 48 Habitat homes

Motivating curiosity and self expression. Independent student projects in language arts.

Science skills
communicating

Concepts
Scientific information may be presented in a number of different formats, including creative ones.

Skills practiced
library research
organizing information
verbal and visual communication

Time
three to four weeks outside of class

Mode of instruction
students may work individually or in pairs or groups

Sample objective
Students communicate scientific information through creative writing.

Introduction

People learn scientific information in many ways, not just from lectures, hands-on activities or textbooks. In this activity students communicate to others about their favorite aquatic habitat through a writing project. They may choose to write to inform, to persuade or to express personal ideas. For example, the television script might inform one about a habitat, persuade one to vote for its protection or tell a story about something that happened there. Alternately, you may assign a specific writing task, depending on the rest of your curricular needs. Give students sufficient time to research their habitat.

Action

Hand out the habitat cards to groups of students to remind them of the aquatic homes they have identified. Each student (or group) should select a habitat that he or she particularly likes. Then choose from one of the following tasks or propose a similar project which you feel is appropriate:

- describe a typical day in the life of one of the animals that lives in your chosen habitat
- write and design a travel brochure that would persuade a person to take a trip to see this habitat
- make up an adventure you might have had in this habitat
- write a poem about how this habitat makes you feel when you are in it
- write a letter to your local or regional politicians describing why they should preserve a local aquatic habitat
- write a review like a theater or movie review of a habitat, rating different aspects of its environment
- write four weather reports for your habitat, describing a typical winter, spring, summer and fall day
- pretend you could shrink to the size of a small fish in your habitat and tell a story about something that might happen to you

- write a script for a 1 minute television piece for your habitat and make a "storyboard"; the piece may be informative, persuasive or narrative
- describe how your school could make an aquatic habitat as a schoolyard environmental project
- write a song about your habitat
- write a short play about an imagined (but possible) event in the lives of the animals in your habitat

Before completing the projects, each student should have at least one other student review a draft of his/her work and comment on it.

Results and reflection

Have students reflect on their use of the writing process in a journal or in class, responding to such questions as:

- What was the purpose of your writing?
- What was the audience you were addressing?
- How did you plan your writing?
- What planning procedures helped you?
- How could you have improved your planning?
- What problems did you have to solve in your writing? How did you solve them?
- How did you know when you were finished?
- What will your writing help people learn about your aquatic habitat?

Provide an opportunity for children to share their work with each other, with other classes or with parents. Publish a literary journal. Perhaps even sell it.

Conclusions

Let the students pick one or two class projects and do them. Produce the play (complete with costumes), sing the songs, recite the poems, tell the stories, make the videotape, build the schoolyard aquatic habitat.

Materials
habitat cards from Activity 2

Preparation
If you have your students for all subjects, it will be easy to fit this activity into your program. If you are a science teacher, you might choose to work with the English teacher and/or art teacher at your grade level to accomplish these projects.

Outline
before class
1. decide on list of options for writing assignment

2. print and copy list for students

during class
1. hand out habitat cards and list

2. review and share projects

3. publish results

ACTIVITY 49 Habitat interpreters: designing an aquarium graphic panel

How do we communicate scientific information? Language arts and art in support of communication of scientific information.

Science skills
library research
communication
organization of information

Concepts
Scientists frequently work together not only in the lab or field, but on written communications.

Skills practiced
graphing
drawing
use of computers

Time
2–3 class periods and independent outside work over several months

Mode of instruction
independent group work

Sample objectives
Students work together to research, write and present a project on an aquatic habitat.

Students practice concise written communication.

Builds on
Activity 2

Teacher's information

Consider making this a joint project between art and science class. This is a group project which combines library research skills with verbal and visual communication. Students can also use electronic media as sources, including on-line information. This project also requires practicing the organizational and leadership skills required for working in groups. In our own classes we have kept groups from coming to blows, but otherwise left it entirely up to the students to learn how to work together on a deadline. Eventually, students have to learn to work together without an adult telling them how.

Introduction

Ask students if they have ever visited an aquarium or zoo. How did the zoo or aquarium communicate information to its visitors? The written information and pictures are called the graphics. A single large "poster" is called a graphic panel. Small labels, called identification labels, help you learn about the individual animals in an exhibit. Challenge your students to research, write, test, illustrate and produce a graphic panel for one of the aquatic habitats from the habitat cards, working in groups. In addition, each student in the group will design and write an identification label for an animal that would live in that habitat.

All information must come from the same kinds of sources a zoo or aquarium staff person would use: books, research papers, magazines or on-line information. Asking another person to do your thinking for you by writing for information is not legal. Libraries, aquariums and zoos get dozens of letters a week from children which essentially ask the staff to write the child's report. The return letter often discusses how to learn something for oneself rather than asking an adult.

Action

Divide the class into groups of at least three or four. You may choose the students with their personal abilities in mind or you may wish to make them work in new groups by having them draw their group number. Each group either selects a habitat or has one assigned to it. They might even draw their habitats randomly. Use the habitat cards from Activity 2 for choices of habitats. Keep in mind that some (mangroves or bogs) might be harder to research than others.

Each group will communicate the results of its library research to the rest of the class by making a large poster or bulletin board which includes pictures, graphical organizers, and written information. The model is an information panel at an aquarium or museum. This will be informative writing in which a premium is placed on precise, accurate, concise writing. Give each group a copy of the rules. Additionally, each group will write a question for their colleagues which can be answered in writing in less than 2 minutes and evaluates general understanding of the habitat graphic panel. This question must have a short written answer; no multiple choice or true false. It should require higher level thinking rather than just recall. For example, the question might require the reader to compare, predict, or draw a logical conclusion from what was read.

Pick a standard size on which students must work, based on space available in your room. You might have them work on posterboard or the folding cardboard designed for science fair projects. They may work at home, but the actual application of their work to the allocated space may have to happen during class time because it should be a group project. The primary objective for the poster or bulletin board is to teach their classmates about their habitat. All the materials should be made by the students' own hands or on the students' or class's computers or with the students' own cameras.

Results and reflection

When the assignments are completed, have the students present their work and let the other students read the posters. Lump the questions into a quiz as evaluation. Discuss in a general way what was best

Materials

for class
posterboard
colored pens or crayons
construction paper
scissors
glue
ruler
habitat cards from Activity 2
optional
computers and printers
camera

Preparation

Copy the rules sheet and gather materials.

Outline

before class
1. copy the rules

2. collect the art supplies

during class
1. reflect on how a zoo or aquarium communicates

2. divide into groups

3. use habitat cards as prompts to select topic

4. research habitat

5. design and produce panel

6. share projects and take "tests"

7. reflect on project

8. reflect on group function

about each project in terms of communication. Effective use of pictures, maps, graphs and tables reduces the number of words used and increases the speed of communication. Have students discuss and evaluate the different sources which they used in finding information. Which were best? Least helpful?

Have the students comment in their journal, describing their own contributions to the project. Have them suggest how they might have improved the work. Perhaps more important, have the students reflect on how well their group functioned. Did they have problems working together? How did they resolve their differences? How would they improve the way the group worked together?

Conclusion

The students could choose to award prizes by voting for several categories such as best use of a graph, best use of pictures, most effective communication of the "feel" of the habitat. Pick the most interesting identification label.

Extension and applications

1. Display the habitat panels and identification labels at a parents' day or evening or invite other classes in. Put them up in the hall for other students to enjoy.

2. Visit a zoo or aquarium. Have the students critique the graphics as a part of the trip. How easy were they to read? Were they well designed? Were they well placed? Did they relate to the displays? Have students write follow-up letters to the institution you visited with their comments.

3. Zoos and aquariums spend a great deal of money on their graphics. Most test "prototypes" with visitors before spending money on the final production. Can your students design tests for their own graphics?

4. Aquariums and zoos often have a specific standard for "readability" of their graphics. Students may find readability tests on their computers. Try applying them to their own panel. How easy was it to read their work?

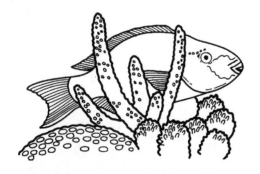

National Aquarium in Baltimore

Rules for the Graphic Panel

Done by one group of students.

- it has fewer than 500 words total, including title, text and captions for pictures
- it is written and designed to be read by fellow students
- it is organized to effectively communicate concepts in order of importance
- it has pictures, written information and at least one graph; maps, charts, and tables are also acceptable
- it covers the characteristics of the habitat as a whole as well as the typical plants and animals found in it
- it addresses one conservation issue specific to this habitat
- everything on it has been made by the students; no magazine cut-outs although photographs they have taken themselves are allowed
- computers and printers may be used to produce text, graphs, etc.
- it will be _____ by _____ in size

bog

Rules for the Identification Label

About one animal from the habitat the group chooses. One label done by each student in the group. The label will be organized in this sequence:

- common name
- scientific name
- where it lives (range) in six words or fewer
- size of average adult or maximum size in metric measurements
- one interesting fact written as a whole sentence or sentences of 15 words or fewer
- a color picture of the animal

The label will be produced on a standard size page (8.5" x 11") with the long side vertical in orientation.

kelp forest

sea grass bed

ACTIVITY 50 An aquatic science symposium

Sharing aquatic science with other students or parents in a scientific meeting and social event.

Science skills
communicating

Concepts
Communication with others is an essential part of the science process.

Time
2–3 classes for organization
1 class period or 1 evening for the event

Mode of instruction
teacher directed group work

Sample objectives
Students organize and run a scientific meeting for peers or parents.

Students communicate the results of their scientific work.

Builds on
entire year

Materials
paper
markers
copier
slide projector and/or VCR

Preparation
Done by students.

Teacher's information

Scientists frequently attend meetings at which they share their most recent research with each other in formal talks, present poster sessions during which they stand beside posters displaying their work and discuss it with interested people, and have informal scientific conversations with people in social settings. Complete your aquatic science unit by holding your own meeting, an Aquatic Science Symposium. The word symposium comes from two Greek words meaning drinking together. In ancient Greece a symposium was a drinking party at which intellectual discussion took place. In modern times the word refers to a conference organized for the discussion of a particular topic.

Introduction

Invite your students to share their hard work. Describe a scientific meeting, building on your experience at meetings. Challenge them to organize one of their own.

Action

Have the students plan the organization of the meeting. Get them to help you list tasks that must be done such as meeting notices (invitations), programs, press releases, refreshments, etc. Invite parents, administrators, supervisors, folks who have helped fund your projects, reporters and/or students from other grades. Use this opportunity to have students report the results of the research from your field trip, demonstrate scientific experiments from the curriculum, display posters of their research projects or give talks, read their creative writing projects and present data from some of the experiments in this curriculum. Issue invitations, decorate the classroom and dream up some aquatic refreshments. Really show off the hard work you and your students have done all year. Do not neglect the social aspects of your meeting. Scientists have a good time at meetings, seeing old friends, catching up on both science and gossip.

GLOSSARY

This is not a list of words to be memorized. It is provided for the teacher's reference. These are words which the students will learn as a part of doing the activities and discussing them in class.

acclimate: an organism changes internal or external features to adjust to changes in the environment; the growing and shedding of a winter coat of heavy hair is an example of a mammalian acclimation process; aquatic organisms generally acclimate by means of internal chemical changes or adjustments; acclimation is a reversible process.

acid: solution in which there are more hydrogen ions than hydroxide ions; has a pH below 7.

adapted/adaptation: an adaptation is a characteristic that was inherited and cannot be changed in an individual; adaptations have a genetic basis and are passed on to offspring.

alga (-ae plural): a photosynthetic organism which lacks the structures of higher plants, such as roots, flowers or seeds; may be single-celled or may be large multicelled organisms such as seaweeds.

alien species: species that have been introduced by humans into new habitats where other species are not adapted to their presence; frequently displace native species.

anadromous: fish that live as adults in the ocean, but swim up into rivers and streams to lay their eggs.

anaerobic respiration: a process that takes place inside cells in which carbohydrate (food) molecules are broken down to release energy without using oxygen; less efficient than aerobic respiration which uses oxygen.

anoxic: without oxygen.

aquatic: growing, living or frequenting water; in this curriculum aquatic includes both fresh and salt water.

atom: the smallest unit into which a chemical element can be broken and still be that element.

base: opposite of an acid, it has more hydroxide ions than hydrogen ions; also chemically reactive; has a pH higher than 7.

benthos/benthic: of or on the bottom of a body of water.

biodiversity: a shortened form of biological diversity used to refer to species richness.

birth rate: births per unit of the population per year; high birth rate supports growth of population size.

bony fish: fish with a skeleton made of bone as opposed to sharks, skates and rays whose skeletons are made of calcified cartilage.

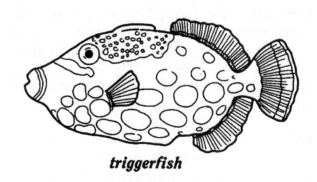

triggerfish

brackish: water which has a salinity between that of fresh water and salt water; a mixture of salt water and fresh water; salinity between 0.5 and 35 parts per thousand.

buffer: neutralizes acids or bases when added to a solution.

buoyancy: the tendency of an object to rise or float when submerged in a fluid; the power of a fluid to exert upward force on an object placed in it.

carnivore: animal that eats other living animals which it catches, as opposed to a scavenger which consumes dead animals.

cartilaginous fish: fish with skeleton made of cartilage, including sharks, skates and rays.

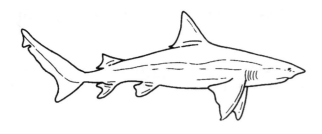

catadromous: fish that live in fresh water and move to the ocean to spawn or reproduce.

characteristic: a feature or trait or property.

classification (-ying): a systematic arrangement into groups or categories on the basis of characteristics shared in common.

cohesion: attraction of molecules of a substance for each other.

cold-blooded: term used to describe animals whose internal temperature is determined by that of their environment; may be quite warm on a hot day. More properly called ectotherms; plants function as ectothermic organisms.

community: an assemblage of organisms living together in association in an environment.

competing: two organisms or populations which need and use the same limited resource are said to be in competition.

condense: in this case, the collection of water molecules in vapor form around a particle or on a surface as they cool and become liquid water.

consumer: an organism that does not do photosynthesis and must feed on other organisms.

control: untreated objects or organisms which are used for comparison against treated objects or organisms in an experiment.

cycle: process in which materials are not lost, but are exchanged continuously among organisms and their environment.

decompose: breaking down of complex organic molecules from living things by bacteria and fungi into the original inorganic molecules from which they were made; decomposers get energy and needed molecules for their own growth.

dense: has a great deal of mass per unit volume.

density: the mass per unit volume of a substance.

diffuse/diffusion: random movement of suspended or dissolved particles from areas of high concentration to areas of lower abundance; mixing until evenly distributed.

displacement: the volume or weight of fluid moved by a floating object.

dissolve: to go into solution; in the case of water, a substance mixes with water and does not settle out upon standing, but stays evenly mixed.

dissolved oxygen: oxygen molecules mixed in solution with water.

ecological balance: relatively stable conditions found in some natural, undisturbed communities over time.

ectotherm: an organism whose body temperature is the same as its environment; some ectotherms use behavior to find a warm place such as a marine iguana warming in the sun; less correctly called cold-blooded.

endotherm: an organism that expends metabolic energy in order to maintain its interior temperature. In cold places the internal temperature is above that of the environment, but energy is also used to keep cool. Mammals and birds are endotherms as are some very large sharks and tuna and leatherback sea turtles; also called warm-blooded.

epineuston: organisms living supported by surface tension at the surface of the water.

erode/erosion: gradual wearing away; in this curriculum water is the agent that causes the wearing away.

estuary: region where salt and fresh water mix in a partially enclosed body of water; generally at a river mouth or in lagoons behind barrier beaches.

euphotic zone: the surface layer of water in an aquatic environment in which photosynthetic organisms can survive; as opposed to the photic zone which is the depth to which any light penetrates and which is deeper.

evaporate/evaporation: liquid becomes a gas and disperses into the atmosphere.

extinct: ceasing to exist; generally used to refer to a species or group of living things that has disappeared permanently.

fisheries management: like wildlife management, an attempt to use wild populations for human purposes without destroying them; requires an understanding of ecological principles and constant monitoring of populations managed.

flagellum (-a plural): hair on a single cell that moves, either moving the cell or moving a fluid over the cell.

flow chart: a visual representation of the choices made in a key

food: chemical compounds containing carbon atoms along with other elements which have been assembled by living organisms and are eaten and used for growth, repair, energy and reproduction.

food web (chain): the sequence of organisms in a community which produce food and consume it; the path that food (materials and stored energy) takes through a group of organisms.

fresh water: water with a salinity of less than 0.5 parts per thousand; no taste of salt.

gas exchange: transfer of gases, principally oxygen and carbon dioxide, between a living thing and its environment.

groundwater: water stored naturally underground, generally in rock layers; comes to the surface naturally in springs. Water reaches these layers by moving down through the soil and rocks from the surface; humans drill wells to reach it.

habitat: place normally occupied by a particular organism; kind of place such as a lake or stream.

halocline: point in an aquatic system stratified with regard to salinity at which the salinity changes rapidly; boundary between two salinities.

hemoglobin: molecule in red blood cells that has a high affinity for oxygen which it carries to sites where it is being used.

herbivore: animal that eats plants or algae (photosynthetic organisms).

ion: an atom or molecule that has lost or gained one or more electrons and has become electrically charged; dissolves readily in water.

indicators: color changes that make visible amounts of invisible products in solution or that change when chemical reactions that cannot be seen take place; markers for things that cannot be seen.

impermeable: cannot be penetrated.

infiltrates: moves between particles into something.

intensity: the amount of something, used here for the amount of light (photons).

larva (-ae plural): immature form of an animal that is physically very different from the adult.

mass: a means of expressing the quantity of a material that is not dependent on gravity as weight is; an object would have the same mass on the earth and the moon, but not the same weight. Within one gravitational field, mass is the same as weight.

migration: movement of animals from one area to another; frequently done on a seasonal basis between specific areas.

mineral nutrient: inorganic molecule needed in the function of living things; usually an essential part of larger molecules or to make enzymes function efficiently.

modeling: process in which one constructs a system which attempts to reproduce aspects of a real system which can then be tested.

molecule: smallest possible unit of a compound substance; has two or more atoms.

nekton: free-swimming aquatic animals that are independent of currents or waves.

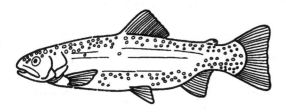

net photosynthesis: the amount of photosynthesis that can be measured by gas exchange; less than actual photosynthesis due to respiration.

neutral: a solution in which the hydrogen ions are equal to the hydroxide ions, making it neither acidic nor basic (alkaline). Expressed as a pH of 7.

neuston: organisms that live at the surface of the water.

nitrogen: an element essential for living things; required to make proteins and nucleotides; frequently in low supply in marine environments where its lack may limit phytoplankton growth.

non-point source pollution: pollution that enters water through runoff from the land.

nutrients: in the case of plants and other photosynthetic organisms, chemicals from the environment which are necessary for life and growth.

omnivore: organism that feeds on both plant and animal sources; feeds at several levels of the food chain.

optimal/optimum: most favorable or best condition for an organism.

organism: any living thing.

oxygen: an element common in the atmosphere as a two atom molecule; a waste product of photosynthesis, it is required by plants and animals for respiration.

parts per million (ppm): a method of expressing the concentration of salts or other materials in solution based on the relative weight of the salts to the solution.

parts per thousand (ppt): a method of expressing the concentration of salts or other materials in solution based on the relative weight of the salts to the solution.

percolate: fluid moving down between particles.

permeable: something that can be penetrated.

pH: a way of expressing the acidity (pH 1–6) or alkalinity (pH 8–14) of a solution by measuring the concentration of hydrogen ions; since it is based on powers of 10, a pH of 1 has 10 times as many hydrogen ions as a pH of 2 and is 10 times more acidic than 2.

phosphorus: an element required by living things; as a plant nutrient it is most commonly limiting in freshwater environments.

photosynthesis: chemical process which takes place inside cells in which light energy is used to make carbohydrates from carbon dioxide and water; oxygen is a waste product of this reaction; done by plants, algae including seaweeds and phytoplankton, and some bacteria.

phytoplankton: small, generally microscopic aquatic organisms that are photosynthetic and drift with the currents; generally single-celled; include many kinds (phyla) of organisms all called algae.

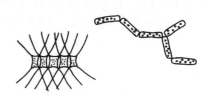

point source pollution: pollution that is released from a specific known source that has an exact location.

precipitation: water falling to earth in the form of rain, sleet, or snow.

predator: animal that hunts and eats other animals.

prey: organism that is eaten by a predator.

producers: organisms that make food; in this curriculum primary producers, generally photosynthetic organisms, are simply called producers.

quality: in this curriculum, the wavelengths of light available.

rate: the speed at which something happens, amount per unit time.

respiration: chemical process which takes place in the cells of plants and animals in which carbohydrates are broken down and energy is released which can be used by the cells to do work; most common form involves the use of oxygen and the release of the waste products carbon dioxide and water.

runoff: water that is not able to percolate (sink in) due to falling on a hard surface; it flows over the surface downhill.

salt: a crystalline compound made up of ions which tend to separate readily in water; hence salts dissolve rapidly.

salinity: saltiness of a solution, expressed as ppt generally.

salt water: ocean or sea water; salinity of 35 parts per thousand.

saturated: something that has as much of something else in it as it can hold under usual conditions.

seasons: spring, summer, fall and winter; different seasons characterized by climatic conditions caused by Earth's rotation. Seasons are not as apparent in tropical areas where there may be wet and dry seasons.

sessile: attached; not free-moving.

simulation: a model system that imitates a real situation.

solution: a mixture, usually liquid, in which one or more substances are distributed throughout the liquid in the form of separate molecules or ions; cannot settle out; appears transparent though may be colored.

spawn: release eggs and sperm for fertilization in the water.

specific heat: the amount of heat a substance absorbs when its temperature increases one degree centigrade; different for different substances.

specific gravity: the density (mass/unit volume) of a substance using water as a standard which is set at 1.0.

spring: point at which the water table meets the surface of the earth and water flows out naturally; also a season.

standing crop: total weight of animals or plants in existence at a point in time in a defined area; as opposed to all the plant or animal material produced over a period of time, such as a year.

stratified: layered.

subtropical: climate in which freezing is an occasional, but not annual event; above the tropics.

surface tension: condition at the surface of a fluid which acts as an elastic film due to the molecular forces within the fluid.

surface water: water on the surface of the ground such as a lake or river.

suspension: a mixture, usually liquid, in which a substance may settle out of the liquid when allowed to stand.

swim bladder: also called an air bladder; gas filled organ found in most bony fish which is inflated or deflated to adjust the buoyancy of the fish, and thus change its position in the water.

temperate: climate in which distinct seasons alternate on an annual basis; both photoperiod and temperature variation that includes routine freezing weather.

thermocline: boundary layer of water where temperature rapidly changes from warm surface water to colder bottom water.

tolerate: endure, resist or survive without grave or lasting injury.

tropical: moderate temperature and photoperiod cycles with no freezing weather; may have pronounced wet and dry seasons.

turbid: prevents the passage of light; cloudy or opaque.

turnover: mixing of surface and bottom water in a previously stratified system.

variable: condition which is subject to change.

ventilation: movement of air in and out of lungs or movement of water over gills.

vertical migration: movement up and down in the water, rather than horizontally; generally done by zooplankton on a daily cycle of up at night and down during the day.

viscosity: resistance to flow.

warm-blooded: old term used to describe animals that regulate their internal temperature to a constant warm temperature; more correctly called endotherms.

warning coloration: bright color or color pattern that advertises an animal as dangerous or poisonous.

water column: term used to describe the vertical dimension in an aquatic habitat.

watershed: all the land that drains into a specific body of water.

water table: top level of groundwater in soil.

water vapor: water in its gas form.

well: hole dug by humans to reach the water table and drinking water.

zooplankton: generally small to microscopic aquatic animals, larvae or eggs that are not strong swimmers and drift with currents; some members such as jellyfish may be quite large; may be a temporary resident of zooplankton or may be a permanent member.

RECIPES

SOURCES OF EQUIPMENT AND ANIMALS

Consult the annual supplement to *Science and Children, Science Scope* or *The Science Teacher* (from the National Science Teachers Association, 1840 Arlington Blvd., Arlington, VA 22201-3000) for a comprehensive list of suppliers for K-12 scientific equipment, media materials and computer software. Many of the materials in *Living in Water* come from the grocery store, dime store and/or hardware store. The rest can be purchased from a large supplier like Carolina Biological Supply, 2700 York Rd., Burlington, NC 27215 (1-800-334-5551). Kits for most of this curriculum are available from Delta Education, P.O. Box 915, Hudson, New Hampshire (1-603-889-8899). Get a supply list from Delta and check it against the materials lists for the activities you wish to do.

Water test kits may be purchased through supply catalogs or direct from LaMotte Chemical Co., P.O. Box 329, Chestertown, MD 21620 (1-800-344-3100). Another test kit supplier is the Hach Co., P.O. Box 389, Loveland, CO 80539 (1-800-227-4224). These companies' kits are different from each other. The parts are not "mix and match."

Field sampling equipment can be ordered from most supply houses as well as LaMotte and Hach. A couple of not so well known sources for collecting gear are Memphis Net and Twine Co. which makes an extensive selection of commercial fishing nets (1-800-238-6380) and Leave Only Bubbles, Inc. which has the best cheap zooplankton net made as well as other field gear and good marine books (1-800-890-0134).

MAKING THINGS IN SOLUTION

Making salt solutions

Do not use salt sold as regular table salt (for salt shakers) to make unknowns. It contains anti-caking additives which make it cloudy. Kosher salt and canning salt do not have these additives. Kosher salt is sold in large boxes; canning salt comes in 5 pound bags.

Ideally, you would mix your salt solutions the way a scientist does: weigh 35 gm of salt for each liter of 35 ppt solution you need. Put 35 gm of salt in the bottom of a 1 liter container and fill to the 1 liter mark with water. For 70 ppt, use 70 gm salt/liter. Since parts per thousand is based on weight, 35 ppt sugar would be made exactly the same way.

Don't have a good balance? Here is a quick and dirty way to accomplish about the same thing for salt (NaCl) solutions only. For 35 ppt put 2 scant tablespoons of salt in 1 quart of water. For one gallon of 35 ppt use 7 level tablespoons of salt. For 70 ppt, just double the salt in the same amount of water.

Making nitrate solutions with Nutritabs

You can make solutions that contain nitrate using any normal commercial fertilizer. Use a nitrate test kit to adjust the concentration. For a short cut, use LaMotte Nutritabs which are designed to give specific concentrations. They are made to be used in LaMotte curriculum materials, but can be ordered separately from LaMotte. They are item #5421, 50/box.

Making aged tap water

Water supplied by most municipal districts is treated with chlorine to kill bacteria which might be in the water. Chlorine is also toxic to most aquatic plants and animals. Consequently, if you put fish or plants in water right out of the tap, they may die. Chlorine is a gas dissolved in the water. If the water is exposed to air for 24 hrs, the chlorine gas will diffuse out of the water. Gases in the water will reach equilibrium with those in the air during this time. Fortunately, there is very little chlorine gas in the air, so the chlorine leaves the water. Always allow water from the tap to sit for at least 24 hrs before animals or plants come in contact with it.

Making fresh water with low and high dissolved oxygen

Fresh water left uncovered in a refrigerator or in an ice chest or outside on a cold, but not freezing day, for 24 hrs will have 10–11 ppm dissolved oxygen. Fresh water left sitting out uncovered at room temperature at about 22–25 $^{\circ}$C for 24 hrs should test at about 8 ppm dissolved oxygen. At each temperature, salt water will have lower dissolved oxygen than fresh due to the salinity.

Heating water drives off all dissolved gases, including oxygen. If you boil water vigorously for 15 minutes and then gently pour it into a canning jar and seal it, the water will have less than 1–2 ppm dissolved oxygen. Canning jar lids have a rubber ring that seals as the jar cools. If you use a jar not intended for canning, such as a mayonnaise jar, it may shatter when the hot water hits it. If you must use a jar other than a real canning jar, put the jar in the sink. Put a table knife in the empty jar. Then pour the boiling water into the jar. It may still shatter, but the sink will catch the flood and glass. Generally, the knife trick works. The jars must be filled completely with no air space at the top. Remove the knife. Seal immediately with a canning jar lid and do not open until just before use. A blunt table knife can be used to pry the jar lids up to open them.

For dissolved oxygen in the range of around 4 ppm, heat water on the stove to about 50 $^{\circ}$C and hold it at that temperature for 1 hr or more. Another, easier way to get the water to hold at the correct temperature is to use a hot tray, heated serving tray, electric skillet or hot pot set at 50 $^{\circ}$C. For best results, allow the water to stand at the desired temperature overnight before canning it. Seal the water in canning jars with canning jar lids. If you do not have a thermometer which reads in centigrade, try a meat thermometer at 120–130 $^{\circ}$F.

Canning jars are widely available in grocery or hardware stores in summer and early fall. Sealed prepared water keeps on a shelf until use. You can make it several months ahead of time. Do not try to use plastic. Some plastics are permeable to gases. Many will wilt with boiling water to give an "interesting" new shape.

For safety and convenience, heat the water at home or in a prep room when students are not around. Allow all heated water to reach room temperature before children use it. If it has been sealed properly, no oxygen will enter until you open the jar, just before use.

MOVING WATER AND TAKING SAMPLES

Pouring

If you have lots of the sample and a sink or pan under the sample, carefully pour the sample down the side of the dissolved oxygen sample bottle, allowing the rest to fall into the sink.

Siphons

Water is easily moved from one container to another using a siphon. Clear, flexible plastic aquarium tubing used for siphoning may be purchased from scientific supply catalogs, but is also sold in hardware and auto supply stores and pet shops featuring fish. Siphons may be started by sucking on one end with the other immersed in the solution to be moved. This practice usually results in a mouthful of the solution. Small plastic tubing clamps which can be used to regulate the flow rate prevent this problem by allowing you to seal the tubing just before the solution reaches your mouth. They are available from supply houses.

The other end of the siphon must be at a level lower than the surface of the solution being siphoned. Remove your finger or open the tubing clamp to allow flow. Because germs will be transmitted if students are allowed to use siphons, make yourself the sole siphoner and wash your tubing regularly. A siphon may also be started by holding the entire length under water until it fills. Remove one end while holding it closed and proceed as above.

Kitchen basters and syringes

Water samples for the dissolved oxygen test kits may also be transferred with a large kitchen baster if the bulb is squashed flat before being immersed in the solution to be sample. No blowing bubbles in the solution to be sampled! The sample is dribbled down the side of the sample bottle, not squirted in from above.

Large syringes with graduated measurements are available from several suppliers of elementary science materials. These are made of plastic and look like giant "shots" except that they lack a needle. With either the kitchen baster or the syringe, there is the chance that they will be used as squirt guns. It is absolutely critical that the water not be squirted into the dissolved oxygen test bottle. Dribble the water down the side of the bottle. This can be accomplished with a syringe by filling it to the point where the plunger is about to pull out of the top, placing the base inside of the top of a tilted sample bottle and pulling the plunger on out the top. The water will trickle down the side of the sample bottle. Use the gentlest possible way to move water so you do not add oxygen.

TEST KITS FOR THINGS IN SOLUTION

Testing for pH

There are dozens of pH kits and test papers. You may substitute test paper in these exercises, but it must be fresh. Its accuracy declines with age and exposure to air. The suppliers listed in the opening "Sources" have a number of pH testing products. A wide range pH test kit is most useful if you are likely to encounter conditions ranging from severe acid mine drainage (pH of 4) to spring runoff from fields recently limed (pH 9.5) for acid soil problems. Expand the value of one kit by ordering extra vials for samples and sharing the chemicals and color comparator among groups. The pH kit chemicals last. There are also electronic pH meters, including equipment that can be hooked up to a computer.

Testing for nitrate

Read the description of the nitrate test kits very carefully. Do not select one that is based on cadmium reduction. Cadmium is a toxic heavy metal. LaMotte has a zinc reduction kit that is accurate and much more appropriate for student use. Zinc is also a heavy metal, but it is an essential mineral in human diets and much less toxic than cadmium. The kits measure nitrogen in the nitrate ion in ppm, the same units that the Environmental Protection Agency uses in regulations. We use the LaMotte kit #3345 in our classes. Order extra sample vials. The chemicals #1 and #2 are individually sealed tablets. Eight groups can be testing a sample, each with its own vial and tablets. Then pass the comparator to read the results. This way you use the tablets from one kit quickly (50 tests per kit) and order fresh ones. Do not use tablets more than a year or two old or that have a break in the seal. Exposure to air also ruins them.

Testing dissolved oxygen
Dissolved oxygen test kits

Dissolved oxygen test kits are available from many companies. Two we use are LaMotte Chemical Co. and Hach Co. kits. The Hach kit contains dry powders. The LaMotte kit has chemicals in solution plus one powder. Both work well. A third company markets simple glass ampoules, but the tests are less accurate and do not generate data with decimals. The LaMotte kit #7414 is the most frequently used by our teachers. It is widely available through biological and scientific supply firms. Materials may also be purchased directly from LaMotte (P.O. Box 329, Chestertown, MD 21620; 1-800-344-3100). *Living in Water* gives instructions for the LaMotte kit.

Making extra kits

Each piece of the LaMotte kit is independently available. Replacing the chemicals or kit parts is easy. Extend the usefulness of the kit by combining a single dissolved oxygen test kit with three extra sample bottles, three vials and three syringes to provide enough equipment for 4 groups to work at the same time at a total cost of about $60, less than the price of 2 kits. The sample bottle is 0688-DO, the syringe is 0377, and the vial is 0299. Ask for both the reagents and the parts (kit) catalogues.

The chemicals from the kit are used one at a time and passed along to the next group, assembly-line style. This heavy use of the chemicals from one kit for four groups assures that all the chemicals will be used during the year, keeping the chemicals fresh. The chemical set does a minimum of 50 tests per chemical refill. Check with other schools to see if anyone has an old kit. Kits 15 years old still used the same parts. New chemicals should be purchased each year. The parts number is written on each piece.

Modifying LaMotte kit #7414 for dissolved oxygen

Detailed instructions are included with the LaMotte dissolved oxygen test kits. The instructions are complicated for children. Read all the original instructions and file them for your own use. Copy the instructions in this curriculum and laminate or cover them with plastic for your students' use. To make the kit match these instructions, label the sample bottles and solutions with tape and a permanent marker as shown below. Do not cover up the original labels with chemical names and safety precautions. Explain to your students that this is an accurate test for dissolved oxygen. When they take chemistry, they will understand how it works. For now, they can take it on faith. It is the same test scientists used for many years before dissolved oxygen electrodes were developed. Label the back of each bottle in the test kit as follows:

Empty sample bottle (0688-DO) with screw cap—A

Empty sample vial (0299) with snap cap with hole in it—B

Manganous sulfate solution—1

Alkaline potassium iodide azide solution (the most toxic chemical)—2

Sulfamic acid powder—3

Starch solution—4

Sodium thiosulfate solution—5

The instructions for use of the LaMotte kit are new in this edition of *Living in Water*. To use kits previously labeled according to the first or second edition of *LIW*, renumber the chemical bottles, reversing chemicals 4 and 5.

Demonstrate the use of the test kit, using students as helpers, for the whole class before any students try it without supervision. Have one student read the instructions while another does the test. Make sure your students are aware of the fact that they should regard all chemicals as dangerous and handle them with care and respect. Avoid all contact with skin or eyes. Never taste any chemical. Wear eye protection whenever using chemicals.

Testing dissolved oxygen in the field

In the field, have the students take the water samples and record information on source, etc. A teacher then adds chemicals 1–3 (steps 1–6) to each sample bottle promptly and seals it. At this point the samples will hold without change and may be taken back to school. This precludes children handling chemicals under uncontrolled field conditions. The final steps are done by the students on a subsequent day at school working carefully at their desks. This requires enough sample bottles for all the sampling done on a field trip. You cannot store untreated water samples for any length of time since bacteria and phyto-plankton in the water may use oxygen before measurements are taken. Fixing the samples in the field solves this problem.

Filling a water sample bottle for a dissolved oxygen test

Rinse the bottle with sample water first if possible. If sampling from a container, you can carefully pour the sample down the side of the screw-capped sample bottle (A). Pouring is the simplest technique. You may also use a kitchen baster or syringe to remove water if it is never squirted or splashed. Avoid the mistake of squeezing the bulb or pushing the plunger under water in the sample. The teacher may siphon water out of the test jar into the sample bottle with aquarium tubing. For safety, siphoning should be only done by the teacher. It is a good way to collect bubble-free samples.

In the field in shallow water, put the sample bottle (A) at the depth you want to sample. Hold it sideways. Remove the cap and fill completely, tipping up at the end. Cap tightly underwater. If getting water from a water sampling device, pour gently down the side of the bottle. Do not splash or make bubbles. LaMotte now makes a water sampler (Code 1054-DO) for field use that holds the DO sample bottle (A) perfectly. Just pull it out and fix the sample.

New Instructions for LaMotte Dissolved Oxygen Test Kit (code 7414)

Danger: poisonous chemicals!
Do not taste or get on skin or in eyes.
Use eye protection when working with chemicals.

This kit should contain: 2 empty bottles marked "A" and "B," 5 chemical bottles marked 1, 2, 3, 4, and 5, an eye dropper (which may be in the cap of chemical 4), a scoop and a syringe. (If the liquid chemical in the brown screw cap bottle is labeled "5," bottles 4 and 5 must be renumbered before you can use these instructions. Have your teacher check the numbering instructions in his/her manual.) Always put the cap back on the chemicals immediately after use! Wear safety goggles. Put newspapers on your desk to catch spills. In addition, you may want to work on a pie plate.

1. Fill Bottle A full to overflowing by pouring the sample you are testing slowly down the side of Sample Bottle A without splashing it or making bubbles in the sample. Pour over the sink or an aluminum pie pan to catch spills. Empty the pan and set the filled sample bottle in it.

2. Add 8 drops of chemical 1 to the water sample in Bottle A. Hold the chemical bottle vertically upside down. The drops will sink.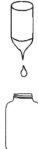

3. Add 8 drops of chemical 2 to the water sample in Bottle A. Again, hold the chemical dropper bottle upside down.

Catch spills in the pie tin. The spills do not contain the added chemicals which sink because they are more dense than water.

4. The sample in Bottle A will change color. Screw the cap on tightly and shake gently for one minute to mix. Then, let the sample sit for 1 minute.

5. Fill the white scoop with chemical 3. The top should be level, but do not use your fingers. Use the inside of the chemical 3 container to level it. Remove the cap from sample Bottle A and add the scoop of chemical 3. Turn it completely upside down over the open top of Bottle A. Do not get the scoop wet. Tap the handle of the scoop on your finger if the powder sticks.

6. Replace the sample cap and shake until the brown flakes go away. This may take a while. If you want, you can stop at this point and the sample will not change as long as the lid is on tight. Fix samples taken in the field to this point and return to classroom for final testing.

7. Pour the colored sample from Bottle A into Bottle B up to the white line. Look at it from the side and make the bottom of the curve of the water surface even with the line.

8. Use the eye dropper to add 8 drops of chemical 4 to the sample in B. It will turn blue. Cap Bottle B and swirl. Do **not** put your finger over the hole and shake. Hold the top in one hand and make the bottom of the bottle go in a circle, causing the solution to circle around in the bottle.

9. Push the plunger on the syringe all the way in. Put the syringe into the hole in the top of chemical 5. Turn the bottle upside down and **slowly** pull the plunger down until the bottom of the plunger is at 0 (see the drawing). Do not pull it all the way out! **Do this slowly.** You may set the syringe on the desk without having it spill.

10. Put the syringe into the hole in the top of Bottle B and add chemical 5 to the sample one drop at a time. Mix by swirling **after each drop**. Do **not** push the syringe hard. Do **not** hold it like a shot. Hold the sides of the plunger with thumb and finger and wiggle it down to get drop. Swirl the sample each time to mix.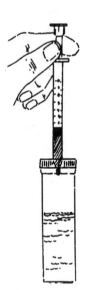

11. When the blue appears to lighten before mixing, **slow down** even more. Continue to add one drop at a time and swirl until the blue just barely goes away. **Stop immediately when the sample turns clear.** If the sample stays blue and the syringe is empty, refill it completely and continue. Add the 10 from the first syringe to the number from the second.

12. When the blue disappears, look at the end of the plunger on the syringe. Read the number where the bottom stopped. Write down the number and write ppm behind it. This stands for parts per million.

This reads 7.4 ppm

13. Dispose of the remaining sample as instructed by your teacher and rinse the two sample bottles with water.

14. **Wash your hands with soap.**

MEASURING AND WEIGHING

Measuring temperature

Small, very inexpensive (about $7–8 dozen), plastic-backed thermometers may be purchased from several suppliers. Because the backing is flexible plastic, the glass will snap easily, creating a safety hazard. You might glue wooden tongue depressors or popsicle sticks to the back of the plastic to make them rigid. Metal-backed thermometers are a bit more expensive, but safer. In both plastic- and metal-backed thermometers, the glass containing the red solution tends to slip up and down against the scale printed on the holder. Let all the thermometers sit out overnight on a desk so that they should read the same. Slip the glass up or down until they do show the same temperature. Use a more expensive and accurate thermometer for an idea of what the correct temperature is. When they all read about the same, attach them permanently to the backing with a big drop of strong, fast drying permanent glue.

Weighing things

Balances

A simple balance with a stick and two pans which sits level when empty is adequate for these activities. There are several kinds made of wood or plastic that sell for about $15. More expensive balances which actually weigh things can also be used, but you will have to teach the students how to use them.

Spring scales

Several kinds of spring scales are commonly sold in science supply catalogs, ranging in price from $6–15 each. The Ohaus 250 gram spring scales work well for the activities in this curriculum. They are accurate and hold up well. They can be adjusted to be set to zero with a pan hanging from them. You will have to teach the students how to interpolate between the numbers on the scale. The scales come without weighing pans. Purchase the smallest aluminum pie tins (about 5 inches across) from the housewares section of the grocery store or use the pans from individual chicken pot pies. Punch three holes equally spaced around the rim. Tie three 2 foot long strings through the holes and pull them up above the pan. Tie an overhand knot in the strings about a foot above the pan with the strings equal length. Then tie another overhand knot about one inch above the first and cut off the remaining string. The pan must be suspended from the top loop top, not through one of the strings below in which case everything will be dumped out when the scale is lifted.

Measuring time

Digital watches that read in seconds are becoming very cheap. Send out a call for old ones that are collecting dust in a drawer because the band broke and a new watch is cheaper than a new band. The lithium batteries are good for 5 years or more whereas the bands break quite fast. Keep an eye out for a sale. Many children have watches that read in seconds and even have functions like alarms and timers. I found that the students in our class were much better at working their digital watches than I was so I delegated all time-keeping. Regardless, the children may require a review of the methods used to subtract one time from another.

Measuring volume

Clear plastic measuring cups are sold in the grocery store housewares section or kitchen supply stores. They have both metric and English markings. Many scientific supply houses offer metric volumetric measuring devices (graduated cylinders) which are made of plastic. Neither of these sorts of items would be acceptable for real scientific research because they cannot be sterilized. They are perfect for classroom use as you will not be using any toxic compounds. They will not break easily and will last for years. The measuring cups are not as accurate as the graduated cylinders, a good lesson about selecting tools. Some new measuring cups nest for storage.

MAKING A MILK JUG SPRINKLER

Use sturdy half gallon plastic milk or apple cider jugs. Screw caps are better than snap-on if you can find them. Heat a small (1–2 inch) smooth finishing nail, holding it with pliers, over the stove or a candle. Use the hot nail to melt a one inch circle of holes (at least 25 holes inside the circle) on the corner opposite the handle at the point where the jug just begins to narrow at the top. Melt three holes near the bottom of the handle itself to let air in. Use a permanent marker to make a line below the holes on the handle and write "fill to this line" below the line. Fill to the line with water and put the cap on. To make the sprinkler work, tip it almost straight upside down and be careful not to have your hand over the air holes on the handle.

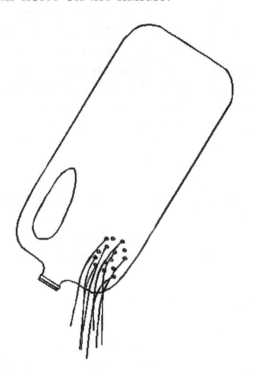

COLLECTING AND LOOKING AT ZOOPLANKTON

While plankton nets can be purchased, you may make a serviceable net from a heavy coat hanger, a sturdy pair of pantyhose and some nylon rope. Bend the coat hanger into a circle. Tie three 3 foot pieces of nylon rope or heavy cord to the rim using strong knots. Stretch the waistband of the pantyhose over the wire and sew it to the wire with sturdy thread. Distribute the rope evenly as you go so that the three ropes are equal distance from each other Fold the band over the wire as you sew for a strong attachment and to avoid runs. The sewn band keeps the ropes in place. Pull the ropes evenly and make an overhand knot in the ends. Use small spice or baby food jars pushed down into the legs about knee level to collect the plankton. Hold them in place with heavy rubber bands. Pull the net along a dock with a rope. Or throw the net out into the water and tow it back. You can also haul buckets of water up and pour them through the net. The net can be pulled by a wader or a snorkeler as well as behind a boat. Attach a fishing weight to the wire for a deeper tow.

Zooplankton can be viewed by the entire class with the deep well projection slides available from Carolina Biological Supply. They are dropped into a 35 mm slide projector. The zooplankton appear to swim around on the screen. These small plastic slides are somewhat expensive, but can be washed and reused. In 1997 Carolina Biological Supply Co. catalog deep well projection slides were item no. K3-60-3730, sold in sheets of 20 for $45. Leave Only Bubbles, Inc. (listed in the opening "Sources") sells individual deep well slides for about $3. This company also has an excellent, inexpensive plankton net.

DISSECTION

Dissection should be used infrequently. Many of us had lots of dissection in our high school and college courses so we continue to use it as a primary mode of lab instruction. It is very useful to see what the insides of a few things look like in terms of understanding them and ourselves, but cutting up dead things is a very minor part of modern biology. This curriculum included one dissection at the insistence of the consulting teachers. Use the fish effectively. Have students work in groups, not pairs.

The best source of fish are absolutely fresh (or fresh frozen) ones from a fisherman or woman you know. Put them on ice immediately upon being caught. It is better to freeze fish, even for a few days, than to keep them refrigerated. The guts go downhill fast. Fresh fish from the grocery store are next best, though they may not be in very good condition inside. Check the insides of one before you buy more. Last choice is preserved fish. The preservative can be somewhat toxic.

A common misconception about dissection is that a scalpel or razor blade is needed. Neither of these items should be used by middle school students. A pair of scissors and some wooden probes such as coffee stirrers are all that is needed. A pair of tweezers is useful. The critical thing is to have the correct scissors: surgical scissors with one sharp point and one rounded point. The sharp point is used to put a small hole in the body wall. The blades are then reversed with the blunt or rounded point inside for cutting along the body wall. Used carefully, very little damage is done to internal organs. Dissections may be done on newspapers or in trays or pans to catch the drips. Small fish may be dissected in Styrofoam meat packing trays. Check with your butcher to see if you can purchase some.

LIVING THINGS

Pond water

A pond water sample should contain small, single-celled algae. Pond water samples or algal cultures can be purchased from a biological supply house. You may find that collecting a natural sample is easy, however. A sample of algae scraped from the sides of a well-established freshwater aquarium may work. During spring, summer and fall samples can be collected from outdoor sources such as a greenish farm pond, city park pond, animal water trough or goldfish pond. Look for sources where the water is rich in nutrients and is standing, not flowing.

Elodea

This cheap common aquarium plant is available from pet stores and biological supply houses. It grows in long strands which tend to float unless anchored. If *Elodea* gets sufficient light, it will live through the entire school year in a classroom aquarium. With care of the plants between use, all of the activities in this curriculum may be done with the same materials. As *Elodea* grows, it will remove animal waste products from the water. Indirect natural light or plant grow lights provide the best spectrum for growth. Aquarium shops may call *Elodea Anacharis*.

Goldfish and guppies

Goldfish and guppies are two of the most commonly available freshwater fish. They are incredibly tough and tolerate the uses in this curriculum quite nicely. Goldfish are somewhat easier to use. They may be purchased from biological supply companies, but are frequently available in pet stores or variety stores at a much lower price. Those sold as "feeder" goldfish are fine. Stick to animals in the 1 to 2 inch size range for the uses here. Feed them flake food. The guppies prefer a warmer tank than the goldfish, so do not mix the two. Use either goldfish (65 ° F) or guppies (75 ° F), but not both. Guppies do fine in a freshwater tropical aquarium with other species. They may surprise you by reproducing. In the case of both species, do not subject them to temperatures above those at which they are kept. **Do not substitute other kinds of fish for use in the experiments in this curriculum.** Goldfish have evolved in low oxygen, high pollution environments.

Brine shrimp

Brine shrimp (*Artemia salina*) are small crustaceans found in salt lakes throughout the world. They are referred to in some places as sea monkeys. They may be purchased as dry eggs which hatch in salt water. The newly hatched animals are truly microscopic. They will not work for these activities. The adults may be purchased live from some pet stores (call ahead of time) and may also be purchased from some biological supply houses. If you are really brave, you might want to try raising the newly hatched animals. Instructions for this project are in the October 1986 *Tropical Fish Hobbyist*. A mother who has raised six generations of brine shrimp for her son's science fair project fish to eat was successful after she learned that the eggs had to dry before they hatched for each new generation. Her animals were reproducing sexually, making eggs that could survive cold and drying. Under other environmental conditions, brine shrimp reproduce asexually, making eggs that develop without fertilization into new adults.

MAKING GRAPHS

There are several different ways of displaying numerical information on a graph. Bar graphs and line graphs are the two most common forms and cover the graphing needs for this curriculum. A bar graph is used when comparing different sets of numbers associated with non-continuous or discrete independent variables such as sugar and salt going into solution. A bar graph might be used to compare the numbers of individuals of different species caught in a net. A line graph is used to express numerical data in which the independent variable (the thing which you changed) is a continuous function, even though you only sampled parts of it. The results (the dependent or response variables) are plotted as points, but the assumption is made that the infinite number of intermediate points possible fall along a line between the points plotted. After plotting the data points you actually measured, you draw a line or curve that best fits those points. An example of a good use of a line graph would be to show the relationship of temperature to dissolved oxygen in water. As temperature increases, dissolved oxygen decreases.

It is desirable to also show the range of results as well as the average to indicate how much variability there was among different groups' results. This can be done in several ways with varying degrees of sophistication. A simple vertical bar through the point or top of the bar graph extending from the lowest value anyone got to the highest value shows the range of data collected. A slightly different tactic is to calculate the median (the middle value) and then make a box and whisker plot around it. From the median, find the median between it and the most distant point on each side. Draw a box around the middle two quartiles with a line through the median. Then draw a line from each end of the box to the most distant data point. These two lines are the whiskers. Consult a mathematics text for more information box and whisker plots.

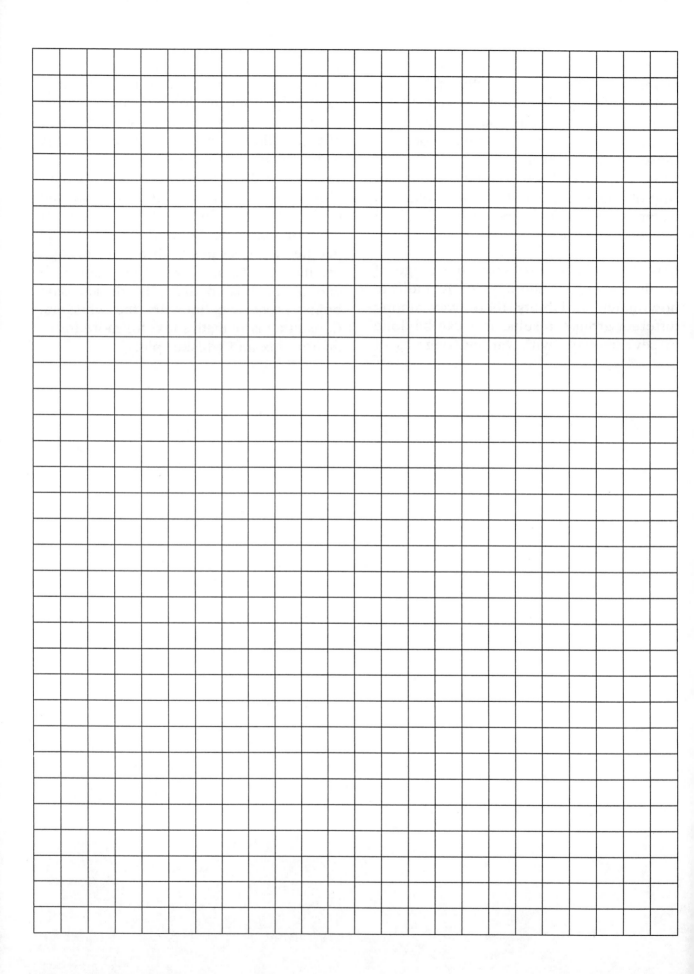

MATERIALS

Group size and managing materials

This materials list is designed to serve a class of 24–32 students, working individually or in pairs, in groups of 3–4 (8 total), in large groups of 8–12 (3–4 total), or in very large groups of 12–16 (2 total). These terms match the descriptions of group size listed under materials in each activity. A small class needs less, a larger class could have five in some groups. We make kits for each group of students with the equipment they use repeatedly. Each group is responsible for cleaning and maintaining its kit. This reduces the materials that must be picked up for each exercise. The kits are stored in the aluminum roasting pans and additional plastic dish pans, one each per group. Student books and papers are moved off the tables during experiments.

We found putting special materials on a rolling cart that we moved into the center of the classroom reduced student travel and traffic. We rearranged the tables to give each group equal access to the materials and a direct line of sight with the instructor by putting them in a half circle with the cart in the center, in front of the instructor's table. The cart rolled to the sink for clean-up and preparation. Passing out papers was also facilitated by this arrangement. Direct eye contact with the teacher improved student behavior and reduced extraneous conversation as there was no back row to hide in.

We assume that students have access to colored pens, markers or pencils for drawing graphs, as well as notebooks or folders for storing papers. Normal pencils make good, waterproof labels.

Reducing waste

Cleaning up after each wet activity has resulted in an over-use of paper towels in our classroom. We should have purchased 2 sponges per group which could be used with our plastic deli containers to mop up with a slightly soapy solution after wet work. Other ways to reduce waste include rinsing the aluminum foil which lasts for many uses.

Purchasing materials

Many of these items come from the grocery store, hardware store or variety store. When we were only buying materials and equipment for our own use, we always shopped locally and saved money. We bought only scientific equipment from suppliers. You may pay twice as much for household items from a kit maker, but scientific equipment is not marked up as much.

When we needed to make workshop kits for almost 70 master teachers, home shopping was no longer an option. Several companies make custom designed science kits for school districts. You need to work closely with them to insure that the materials supplied to teachers are the correct ones. One district testing *Living in Water* for adoption had kits in which the materials did not match the descriptions in the text. For example, 1 inch hose was supplied instead of aquarium tubing and roofing nails were provided in place of finishing nails. Baltimore City Public Schools worked with the authors and a supply company for several months to get exactly the right materials in their *Living in Water* kits. Another Maryland district, Howard County Public Schools, hired a special person to make and stock science kits for their programs. If you are planning on using kits for *Living in Water*, please make sure that the kit materials have been reviewed and approved by the authors (email us at vchase@aqua.org to check). We do not charge supply companies for review as we want you to have the best materials possible.

Substituting materials

There are some things that work better than others. For example, Solo brand flexible plastic cups last forever. Rigid plastic "wine cups" of the same size that look more elegant crack after a few uses. The Ohaus spring scales are cheap, accurate and tough. There may be other materials that are just as good as the brands we list here, but we have not found them yet.

Descriptions

Materials have quantities indicated which are approximately enough to do all the activities listed. The number of each activity which uses this item or material follows the item. A copy of these pages can be used as a shopping list. While household items can be purchased in kits, many are trash: aluminum pie pans, deli containers, plastic milk jugs.

MATERIALS LIST

CHEMICALS AND TEST KITS

household items

- [ ] kosher or canning salt (5 lbs) 1, 2, 7, 8, 9, 10, 29, 46

- [ ] table sugar (1 lb) 1

- [ ] corn starch (1 lb) 1

- [ ] 9 packs dark Kool-Aid 1, 12

- [ ] 4 sets of 4 color food coloring in dropper bottles (16 total) 6, 7, 8, 9, 12, 25

- [ ] 1 pt white vinegar 15, 46

- [ ] 1/2 pt dishwashing detergent 32, 46

- [ ] 1/2 pt cooking oil 21, 46

- [ ] 1 gal distilled water 17, 19

- [ ] variety of bottled water 18

- [ ] dirt with small particle size (clay) 16, 41

- [ ] 20 lbs aquarium pea gravel 11, 12

optional

- [ ] 1/2 cup lemon juice 15

- [ ] 1/2 cup orange juice 15

- [ ] 1 cup rock salt 1

- [ ] 1 cup super fine sugar 1

- [ ] 1 cup sugar cubes 1

scientific supply company items

- [ ] pH test kit 4, 15

- [ ] 6–8 extra pH test kit vials 15

- [ ] nitrate test kit 16, 17, 18, 19

- [ ] 6–8 extra nitrate test kit test tubes 16, 17, 18, 19

- [ ] 2 nitrate test kit replacement chemical sets 16, 17, 18, 19

- [ ] small set (50) Nutritabs 15, 16, 17, 19, 46

- [ ] 2 dissolved oxygen test kits 14, 21, 22, 23, 28, 40, 41, 46

- [ ] 6 dissolved oxygen sample bottles 14, 21, 22, 23, 28, 40, 41, 46

- [ ] 6 dissolved oxygen test kit vials 14, 21, 22, 23, 28, 40, 41, 46

- [ ] 6 dissolved oxygen test kit syringes (titrators) 14, 21, 22, 23, 28, 40, 41, 46

- [ ] 2 dissolved oxygen test kit replacement chemical sets 14, 21, 22, 23, 28, 40, 41, 46

- [ ] 32 safety goggles or glasses 4, 14, 15, 16, 17, 18, 19, 21, 22, 23, 28, 40, 41, 46

MEASURING AND DISSECTING ITEMS

household items

- [ ] 8 sets of measuring spoons 1, 46

- [ ] 8 graduated 2 cup measuring cups (can use 16) 1, 6, 9, 10, 15, 25, 35, 46

- [ ] 16 rulers 3, 13, 21, 22, 35, 40, 41, 46

- [ ] kitchen thermometer 23

- [ ] 4–16 wristwatches or stopwatches (measures seconds) 31, 36

- [ ] 8 calculators 27

- [ ] 8 hanging pans for scales (string and small pie pan) 6, 9, 21, 25, 46

- [ ] 12 ft or longer carpenter's measuring tape 36

scientific supply company

- [ ] 8 10 ml graduated cylinders 15, 16, 46

- [ ] 8 25 ml graduated cylinders 6, 9, 25, 46

- [ ] 8 100 ml graduated cylinders 6, 9, 25, 46

- [ ] 24 thermometers 20, 22, 25, 46

- [ ] weather thermometer 4

- [ ] rain gauge 4

- [ ] 8 simple balances 6, 9, 25

- [ ] 8 250 gram Ohaus spring scales 6, 9, 21, 25, 29, 46

- [ ] 8 dissecting scissors 34

- [ ] 20x–40x power microscope or dissecting scope 34

- [ ] simple prism 37

optional
- [ ] deep well projection slides 31

CONTAINERS AND EQUIPMENT

household items

- [ ] 72 6–9 oz flexible clear plastic cups 1, 6, 9, 11, 12, 14, 15, 16, 19, 25, 32

- [ ] 32 16 oz clear plastic cups 9, 29

- [ ] 8 2–3 oz plastic cups 10

- [ ] 16 12 oz tall jelly (canning) jars preferred for 14, 15, 16, 17, 21, 22, 23, 28, 40, 41, 46 and needed for 20

- [ ] 16 pt or 1/2 pt tall jelly or canning jars 14, 15, 16, 17, 21, 22, 23, 28, 40, 41, 46

- [ ] 8 qt canning jars 1, 20, 21, 24

- [ ] canning jar lids to fit above jars 14, 15, 16, 17, 20, 21, 22, 23, 28, 40, 41, 46

- [ ] 8 1/2 gal plastic milk jugs 1, 6, 7, 8, 9, 17, 19, 25, 29, 32

- [ ] 8 1/2 gal plastic milk jugs made into sprinklers 11, 12, 16

- [ ] 8 1/2 gal plastic milk jugs with section of top removed 35

- [ ] 8 1 gal plastic milk jugs with section of the top removed 35

- [ ] 8 large (1/2–1 gal) deli containers 32, 33

- [ ] 8 small (1/2–1 qt) clear plastic deli containers 10, 11, 12

- [ ] 32 film opaque canisters 29, 39

- [ ] 8 aluminum turkey roasting pans or plastic cat litter boxes 11, 12, 35

- [ ] 16 aluminum pie pans 14, 21, 22, 23, 28, 29, 32, 40, 41, 46

- [ ] 4 plastic buckets 4, 29, 31, 32, 34, 41

- [ ] 32 plastic spoons (soup size are best) 1, 6, 9, 19, 25, 32

- [ ] table knife 23

- [ ] 8 hand lotion or soft soap pump mechanisms 11, 12
- [ ] 100 straight pins 39
- [ ] 150 small paper clips 32, 39
- [ ] 75 small (1–1 1/2 in) finishing nails 35, 39
- [ ] 24 large (2–3 in) finishing nails 35
- [ ] 400 pennies 29, 32

optional
- [ ] 200 florist's glass balls (marbles) 32
- [ ] ceramic coffee mug 6

scientific supply company
- [ ] 2 clear plastic 1–2 gal aquariums or large plastic jars 7, 8, 41, 46
- [ ] 10 ft aquarium tubing 7, 8
- [ ] hose clamp for tubing 7, 8

optional
- [ ] 8 turkey basters or syringes 14, 21, 22, 23, 28, 35, 40, 41, 46

CONSUMABLE ITEMS

household items
- [ ] 200 post-it labels 1, 6, 15, 16, 17, 19, 28, 40, 46
- [ ] 96 bathroom cups 2
- [ ] 8 2 liter clear plastic soda bottles
- [ ] 50 cotton swabs 15, 32
- [ ] 16 3x5 inch white cards 16

- [ ] 16 wooden sticks/pencils 16
- [ ] 8 coffee filters or filter papers 16
- [ ] 120 ft of string 3, 10
- [ ] 32 ft heavy duty aluminum foil 10, 11, 12
- [ ] 40 ft masking tape 10, 35
- [ ] 32 ft clear plastic tape 38
- [ ] newspaper 34
- [ ] paper towels 34
- [ ] balloons 34, 35
- [ ] 1–2 safety pins 34
- [ ] natural history, scuba and conservation magazines 2, 29, 33, 37, 43

ART AND STATIONERY SUPPLIES
- [ ] colored xerox paper 27
- [ ] general art supplies 34, 48, 49
- [ ] 96 sheets 9x12 inch mixed color construction paper 2, 29, 38
- [ ] 16 sheets 9x12 inch red construction paper 38
- [ ] 8 sheets 18x24 inch mixed color construction paper 27
- [ ] 2 boxes plastic modeling clay in 5 inch sticks (4/box) 30
- [ ] butcher/bulletin board paper 26, 29
- [ ] 10 ft 24 in wide blue art supply cellophane 37, 38

- ☐ craft junk—clay, nuts, vials, nails, toothpicks, wire, string, pipe cleaners, styrofoam, film cans, foil, coffee stirrers, straws, glue, etc. 31

- ☐ 100 very fine florist's wires 33

- ☐ local, regional, state and national maps 3, 13, 18

Other materials

- ☐ rubber bands 29, 33

- ☐ 500 poker chips or plastic counters 27, 42, 43, 44

- ☐ 32 sandwich bags 42, 43, 44

- ☐ 3 colors plastic or color flagging 43, 44

- ☐ 8 dice 27

- ☐ 2–3 ft cheese cloth 40

OTHER EQUIPMENT

household

- ☐ ice chest or refrigerator 20, 22, 46

- ☐ ice 10, 20, 22, 46

- ☐ flashlight/slide projector 16, 37

- ☐ large trash can (30 gal) 31, 35

- ☐ ping-pong ball 29

- ☐ camera with flash 15, 17, 46

school

- ☐ whistle 42, 43, 44

- ☐ 16 clipboards 36, 42, 43, 44

- ☐ 4–8 staplers 38

- ☐ 16–32 scissors 38

- ☐ hot light source or direct sunlight 10

- ☐ cool light source or indirect sunlight 15, 17, 40, 41, 46

LIVING MATERIALS

- ☐ 2 goldfish 1 inch long 8, 21, 24

- ☐ brine shrimp 8

- ☐ 2 qt algae culture 15, 17

- ☐ 120 inches *Elodea* 21, 22, 40, 41, 46

- ☐ 4–8 fresh or frozen fish not cleaned or scaled 34